Aeman Saad Mohammed

On Bit Interleaved Space Time Coded Modulation

Aeman Saad Mohammed

On Bit Interleaved Space Time Coded Modulation

Multidimensional Labeling and Iterative Decoding

Südwestdeutscher Verlag für Hochschulschriften

Impressum/Imprint (nur für Deutschland/ only for Germany)
Bibliografische Information der Deutschen Nationalbibliothek: Die Deutsche Nationalbibliothek
verzeichnet diese Publikation in der Deutschen Nationalbibliografie; detaillierte bibliografische
Daten sind im Internet über http://dnb.d-nb.de abrufbar.

Alle in diesem Buch genannten Marken und Produktnamen unterliegen warenzeichen-, marken-
oder patentrechtlichem Schutz bzw. sind Warenzeichen oder eingetragene Warenzeichen der
jeweiligen Inhaber. Die Wiedergabe von Marken, Produktnamen, Gebrauchsnamen,
Handelsnamen, Warenbezeichnungen u.s.w. in diesem Werk berechtigt auch ohne besondere
Kennzeichnung nicht zu der Annahme, dass solche Namen im Sinne der Warenzeichen- und
Markenschutzgesetzgebung als frei zu betrachten wären und daher von jedermann benutzt
werden dürften.

Verlag: Südwestdeutscher Verlag für Hochschulschriften Aktiengesellschaft & Co. KG
Dudweiler Landstr. 99, 66123 Saarbrücken, Deutschland
Telefon +49 681 37 20 271-1, Telefax +49 681 37 20 271-0
Email: info@svh-verlag.de
Zugl.: Ulm, Uni. , Diss. , 2009

Herstellung in Deutschland:
Schaltungsdienst Lange o.H.G., Berlin
Books on Demand GmbH, Norderstedt
Reha GmbH, Saarbrücken
Amazon Distribution GmbH, Leipzig
ISBN: 978-3-8381-1253-4

Imprint (only for USA, GB)
Bibliographic information published by the Deutsche Nationalbibliothek: The Deutsche
Nationalbibliothek lists this publication in the Deutsche Nationalbibliografie; detailed
bibliographic data are available in the Internet at http://dnb.d-nb.de.

Any brand names and product names mentioned in this book are subject to trademark, brand
or patent protection and are trademarks or registered trademarks of their respective holders.
The use of brand names, product names, common names, trade names, product descriptions
etc. even without a particular marking in this works is in no way to be construed to mean that
such names may be regarded as unrestricted in respect of trademark and brand protection
legislation and could thus be used by anyone.

Publisher: Südwestdeutscher Verlag für Hochschulschriften Aktiengesellschaft & Co. KG
Dudweiler Landstr. 99, 66123 Saarbrücken, Germany
Phone +49 681 37 20 271-1, Fax +49 681 37 20 271-0
Email: info@svh-verlag.de

Printed in the U.S.A.
Printed in the U.K. by (see last page)
ISBN: 978-3-8381-1253-4

Copyright © 2010 by the author and Südwestdeutscher Verlag für Hochschulschriften
Aktiengesellschaft & Co. KG and licensors
All rights reserved. Saarbrücken 2010

Acknowledgements

Praise to ALLAH, the most gracious and the most merciful. Without His blessing and guidance, my accomplishments would have never been possible.

I would like to thank all people who have contributed differently to the completion of this thesis. My special thanks, deep gratitude and appreciation to my advisor Prof. Dr.-Ing. Martin Bossert for providing great guidance throughout the period of this study. It was he who helped me enter the world of channel coding. His unlimited encouragement and patience kept me focused on this work. I appreciate specially his human style as the research group leader. I would like to thank Prof. Sergo Shavgulidze, Georgian Technical University, Tiblisi, for his interest in my work and the fruitful discussions during this work.

Then, I would like to thank the colleagues and friends at the Department of Telecommunications and Applied Information Theory for the pleasant atmosphere. Special thanks to Dr.-Ing. Paul Lusina, Dr.-Ing. Stefan Kempf and Dr.-Ing. Bernd Baumgartner for their cooperation and help.

I am very thankful to Dr. Abdulwahab A. Rahim and Eng. Ali Alsoholi for their continuous personal encouragement and support during the period of this study. Thanks to my family. Their support and patience helped me a lot.

On Bit Interleaved Space Time Coded Modulation

Abstract

The coding for the wireless channel is the main topic of this thesis. Mainly, single user transmission over one or more antennas is considered in the first part, where we considered multidimensional Bit Interleaved Coded Modulation with Iterative Decoding using 8-PSK constellations. We showed that an optimum Multidimensional labeling with a designed interleaver outperforms the two dimensional Bit Interleaved Coded Modulation with Iterative Decoding in the whole SNR region when modulation doping is used to compensate for the loss at the low SNR regions. In addition to this, a new interleaver design is introduced. Then we consider the two transmit antennas case and we propose multidimensional constellation labeling for bit interleaved space time coded modulation with iterative decoding using the Alamouti scheme and one receive antenna. The labeling of two 16-QAM signals are designed jointly and optimized using the Reactive Tabu Search algorithm with a slight modification in the fitness function of the two dimensional labeling. The proposed multidimensional labeling provided a large coding gain compared to the best known two dimensional labeling. For the case of two transmit antenna we consider the transmission over two uncorrelated frequency bands and construct a simple $2 \times 2 \times 2$ full-rate full-diversity space time frequency code based on constellation rotation. In the second part the multiuser scenario from information theoretic point of view is discussed.

Contents

1 Introduction & Motivation — 1
  1.1 Digital Communication Systems — 1
  1.2 The Wireless Channel — 2
  1.3 Background & Literature Review — 4
    1.3.1 The Turbo Principle — 4
    1.3.2 Space Time Codes and MIMO Transmission — 5
    1.3.3 Coded Modulation — 6
  1.4 Thesis Outline and Novel Contributions — 7

2 Iterative Decoding, Turbo Codes and Orthogonal Space-Time Block Codes — 9
  2.1 Iterative Decoding and Turbo Codes — 9
  2.2 Orthogonal Space-Time Block Codes — 12
    2.2.1 Encoding of the Alamouti Scheme — 13
    2.2.2 Channel Model — 13
    2.2.3 Detection of the Alamouti Scheme — 15

3 Bit Interleaved Coded Modulation with Iterative Decoding (BICM-ID) — 17
  3.1 Bit Interleaved Coded Modulation (BICM) — 17
  3.2 Bit Interleaved Coded Modulation with Iterative Decoding (BICM-ID) — 20
    3.2.1 The Error Bound of BICM-ID System — 22
    3.2.2 Labeling for Different Modulation Formats — 23
    3.2.3 Labeling of BICM-ID as Quadratic Assignment Problem — 29
    3.2.4 EXIT-chart Analysis — 30
    3.2.5 Interleaving — 31

4 Multidimensional BICM-ID — 33
  4.1 MD-BICM-ID Transmission System — 33
  4.2 Reactive Tabu Search Algorithm — 34
  4.3 MD-BICM-ID with Modulation Doping — 38
  4.4 Interleaver Design for MD-BICM-ID — 38
  4.5 Simulation of the MD-BICM-ID — 40

5 Space Time Signaling — 44
5.1 A Simple Full-Rate Full-Diversity Space Time Frequency Transmission Scheme — 44
5.1.1 The STF detector — 45
5.1.2 Optimum Rotations — 47
5.1.3 Simulation Results — 48
5.2 Multidimensional BISTCM-ID Transmission — 48
5.2.1 Multidimensional Constellation Labeling for the BISTCM-ID — 51
5.2.2 Simulation Results — 52

6 The Multiuser Downlink Scenario (Coding Perspective) — 55
6.1 Comparison of Broadcast and FDMA — 56
6.2 Mutual Information and Channel Coding — 57

7 Summary and Conclusions — 60

Bibliography — 62

List of Figures

1.1	The block diagram of coded communication system	2
1.2	Sketch of multiple reflections	3
1.3	Time diversity	3
1.4	Frequency diversity	4
1.5	Transmit space diversity	4
2.1	Serial and parallel concatenated coding schemes	10
2.2	Iterative decoder for serially concatenated coding scheme	11
2.3	Input output relationship of the SISO module	11
2.4	Alamouti Scheme Model with 2 Transmit and 1 Receive Antennas	14
3.1	The block diagram of a BICM system	18
3.2	Bit metric calculation for BICM	19
3.3	The block diagram of BICM-ID receiver with soft-decision feedback	21
3.4	Subset partitions of 8-PSK for three different labeling schemes	24
3.5	8-PSK channel is converted into four binary channels, each selected by the other two ideal feedback bits	25
3.6	Subset partitions of 16-QAM for four different labeling schemes	26
3.7	16-QAM channel is converted into eight binary channels, each selected by the other three ideal feedback bits	27
3.8	Iterative decoding of serially concatenated codes	30
4.1	2D signal mapping versus 4D signal mapping	34
4.2	4-dimensional 8-PSK signal constellation	37
4.3	Doping technique for MD-BICM-ID with 8-PSK modulation	38
4.4	Interleaver design for 4D BICM-ID with M-ary ($M = 2^m$) modulation, interleaver size is N bits	39
4.5	Performance comparison between 2D, 4D BICM-ID and 4D BICM-ID with doping over Rayleigh fading channel. 8-state, R=2/3 convolutional code, 8-PSK modulation, 4000 information bits/block and 8 iterations are used.	40
4.6	EXIT chart of 4D BICM-ID with 8-PSK modulation over Rayleigh fading channel. $E_b/N_0 = 6.5$ and 8 dB	41
4.7	Performance comparison between different doping ratios of 4D BICM-ID over Rayleigh fading channel.	42
4.8	EXIT chart of MD-BICM-ID with doping technique at $E_b/N_0 = 7.5dB$	42

List of Figures

4.9	BER performance comparison with random interleaver and designed interleaver.	43
5.1	The STF transmitter.	45
5.2	The STF detector.	47
5.3	Performance for different values of θ at SNR=10dB.	48
5.4	The proposed STF code compared to the Alamouti scheme and a STF code based on complex precoders.	49
5.5	The Block Diagram of BISTCM-ID System.	50
5.6	Transformation of two dimensional 16-QAM constellation space into multidimensional 16-QAM constellation space.	51
5.7	Learning curve of RTS algorithm.	53
5.8	BER performance of multidimensional labeling versus two dimensional labeling.	53
5.9	EXIT chart of BISTCM-ID system at $E_b/N_0 = 6$ dB.	54
6.1	Broadcast Transmission (Downlink).	56
6.2	FDMA/Broadcast Channels.	57
6.3	Bit error rates for FDMA, FDMA (known channel) and broadcast.	58
6.4	Bit error rates for FDMA and broadcast.	59
6.5	Mutual information for FDMA and broadcast.	59

Chapter 1
Introduction & Motivation

Wireless communications is enjoying its fastest growth period in history, due to enabling technologies which permits widespread deployment. It follows down a path which began with Hertz and Marconi experimenting with radio transmission in the late 19th century and continues today with an explosion of mobile communications products. These products are mainly concerned with the use of technology to enhance the speed and the efficiency of the transfer of information. Information theory, developed by Claude E. Shannon in 1948 [1], is the key stone to efficient information transmission. It defines the notion of channel capacity and provides a mathematical model by which one can compute the maximal amount of information that can be carried by a channel.

Due to the increasing cost of the bandwidth, spectral efficiency is becoming the most important design parameter in wireless systems. Wireless channels are usually characterized by large attenuation and vagaries in the channel termed as fading. There are other transmission impairments associated with the wireless channels like doppler shift, background noise which together with fading poses a natural hurdle in achieving high data rates. The situation is further complicated due to randomly transmission and geographically separated users. Two main technologies, introduced in the last decades, make it possible to design wireless systems with very high spectral efficiency. The first is the *turbo principle* which allows the design of channel codes that perform near the shannon capacity with low decoding complexity. The second is the *Multiple Input Multiple Output* (MIMO) Transmission which results in dramatic increase in the shannon capacity of the wireless channel compared to single antenna transmission.

The coding for the wireless channel is the main topic of this thesis. Mainly, single user transmission over one or more antennas will be considered in the first part. In the second part a note on the multiuser scenario from an information theoretic point of view will be discussed.

In this introduction basics of wireless information transmission and the motivations of this work will be presented. In section 1.1, basic elements of digital communication systems are briefly presented. The transmission over wireless channel is considered in section 1.2. In section 1.3 a background to the main problems addressed in this thesis will be introduced. We summarize the main contributions of this thesis in section 1.4.

1.1 Digital Communication Systems

Basically, a typical digital communication system consists of three major components : a *transmitter*, a *communication channel*, and a *receiver*. The transmitter translates the information bits into the signals that can be effectively transmitted over the channel. The communication channel is the physical medium where the actual communication takes place. The receiver tries to retrieve the transmitted information bits as correctly as possible.

The use of *channel encoder* at the transmitter of digital communication system results in coded communication system, as illustrated in Figure 1.1. The purpose of the channel encoder at the transmitter side is to introduce, in a controlled manner, some redundancies in the binary information sequence so that at the receiver they can be used to overcome the effects of noise and interference encountered during the transmission of the signals through the channel. The number of information bits divided by the total number of bits at the encoder output is known as the *code rate*. The bit error probability can be made arbitrarily small if the code rate is smaller than the channel capacity. The *modulator* produces a Radio Frequency (RF) carrier representation to the binary information. The simplest modulation considered is Binary Phase Shift Keying (BPSK). In this scheme during every bit duration, one of two phases of the carrier is transmitted.

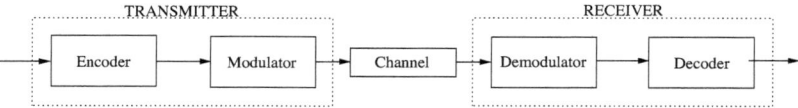

Figure 1.1: The block diagram of coded communication system

The purpose of the *channel decoder* is to find the codeword closest (in some sense) to the received sequence. For most codes the decoding complexity is very large if they do not have some structure that makes the decoding less complex. For example the Reed-Solomon (RS) codes used in Compact Disc (CD) players have as many as $(2^8)^{28} = 2^{224} = 10^{67}$ codewords. Thus comparing the received vector with all possible codewords is not practical. Thank their algebraic structure, the decoding complexity of RS-codes is very low.

1.2 The Wireless Channel

The key characteristics of the wireless channel are fading and multipath propagation. Fading refers to the rapid fluctuation of signal strength over a short travel distance or period of time. Fading is primarily caused by multipath propagation of the transmitted signal, which creates replicas of the transmitted signal that arrive at the receiver with different delays as shown in Figure 1.2. These versions of the transmitted signal combine either constructively or destructively at the receiver resulting in fluctuation in amplitude and phase of the resultant signal.

There are other factors that influence the fading such as speed of the mobile, speed of the surrounding objects and the transmission bandwidth of the signal. During severe fading, the transmitted signal cannot be determined by the receiver unless some less attenuated version of it is available at the receiver. This usually can be achieved by introducing some sort of diversity in the transmitted signal. The three most common diversity techniques are:

Temporal Diversity: Figure 1.3 shows a time diversity scheme at its simplest form. Signal $s(t)$ and its delayed version $s(t + \Delta t)$ are transmitted using the same antenna and the same bandwidth. If both versions of the code undergo different fading then time diversity is achieved. An example of temporal diversity is channel coding with time interleaving. The receiver is provided with several versions of the transmitted signal as redundancy in the temporal domain.

Frequency Diversity: This type of diversity is based on the phenomenon that the structure of multipath propagation depends on the frequency of the transmitted wave. Thus redundancy in the frequency

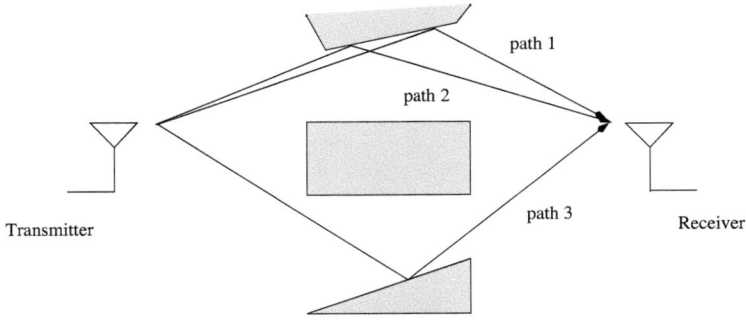

Figure 1.2: Sketch of multiple reflections

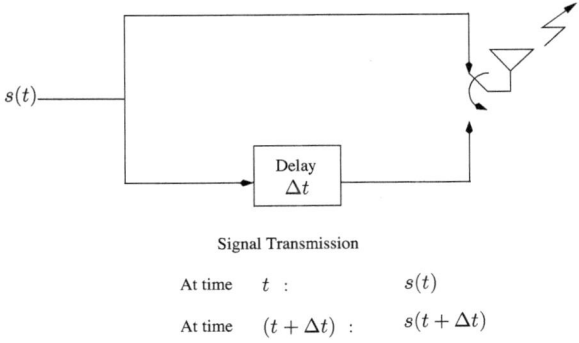

Figure 1.3: Time diversity

domain provides the receiver with several replicas of the transmitted signal that experience different fading at any particular time instance. To ensure independent fading employing this technique, the difference between the carriers, w_1 and w_2 must be greater than the coherence bandwidth, B_{ch}.

Antenna or Space Diversity: In order to create space diversity, several spatially separated or differentially polarized antennas are employed. This would generate redundancy of the transmitted signal in the spatial domain, where each replica would undergo different propagation path. In this context, diversity order refers to the number of uncorrelated spatial branches available at the transmitter or receiver, where the probability of losing a signal decreases exponentially with increasing diversity order.

It is always desirable to employ all forms of diversity in order to combat the adverse effects of the wireless channel. However, it becomes sometimes impractical to employ a particular type of diversity in a specific situation. For example, temporal diversity is ineffective in slow fading channels especially for delay-sensitive applications. In addition, antenna diversity at the mobile unit induces design impracticality. The most common systems that employ different types of diversity techniques for the sake of

1 Introduction & Motivation

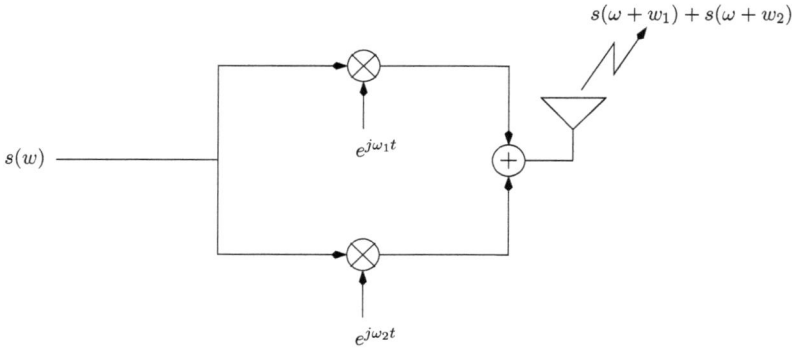

Figure 1.4: Frequency diversity

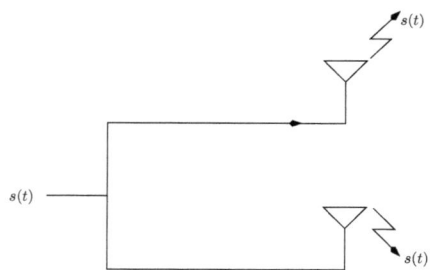

Figure 1.5: Transmit space diversity

improving the performance of wireless transmission/reception are STC and MIMO schemes.

1.3 Background & Literature Review

1.3.1 The Turbo Principle

Concatenated coding schemes were first proposed by Forney [11] as a method for achieving large coding gains by combining two or more relatively simple building component codes. The resulting codes had the error-correction capability of much longer codes, and they were endowed with a structure that permitted relatively easy to moderate decoding complexity. A serial concatenation of codes is most often used for power-limited systems such as transmitters on deep-space probes. The most popular of these schemes consists of a Reed-Solomon outer code followed by a convolutional inner code [12]. A turbo code can be thought of as a refinement of the concatenated encoding structure plus an iterative algorithm for decoding the associated code sequence. Turbo codes were first introduced in 1993 by Berrou, Glavieux, and Thitimajshima, and reported in [39], where a scheme is described that achieves a bit-error probability of

10^{-5} using a rate 1/2 code over an additive white Gaussian noise (AWGN) channel and BPSK modulation at an $Eb/N0$ of $0.7dB$. The codes are constructed by using two or more component codes on different interleaved versions of the same information sequence. Whereas, for conventional codes, the final step at the decoder yields hard-decision decoded bits (or, more generally, decoded symbols), for a concatenated scheme such as a turbo code to work properly, the decoding algorithm should not limit itself to passing hard decisions among the decoders. To best exploit the information learned from each decoder, the decoding algorithm must effect an exchange of soft decisions rather than hard decisions. In a typical communications receiver, a demodulator is often designed to produce soft decisions, which are then transferred to a decoder. The error-performance improvement of systems utilizing such soft decisions compared to hard decisions are typically approximated as 2 dB in AWGN. Such a decoder could be called a soft input/hard output decoder, because the final decoding process out of the decoder must terminate in bits (hard decisions). With turbo codes, where two or more component codes are used, and decoding involves feeding outputs from one decoder to the inputs of other decoders in an iterative fashion, a hard-output decoder would not be suitable. That is because hard decisions into a decoder degrades system performance (compared to soft decisions). Hence, what is needed for the decoding of turbo codes is a soft input/soft output decoder. For a system with two component codes soft decisions are passed from the output of one decoder to the input of the other decoder, this process iterates several times so as to produce more reliable decisions. It turned out that the method applied for these parallel concatenated codes is much more general. Strictly speaking there is nothing turbo in the codes. Only the decoder uses a turbo feedback and the method is known as the Turbo-Principle, because it is applied to many detection/decoding problems such as serial concatenation, equalization, coded modulation, multiuser detection,joint source and channel decoding and others [5].

1.3.2 Space Time Codes and MIMO Transmission

The traditional approach to achieve diversity in wireless transmission is to use multiple antennas at the receiver, so that the receiver has multiple observations that are not deterministically related. The multiple antenna receiver combines the received observation from different paths, and a commonly used method to do this combining is referred to as maximum ratio combining (MRC). In MRC, the received signals at the outputs of the antennas are combined linearly, and the coefficients of the linear combination selected to maximize the instantaneous SNR.With the demand to support reliable, high data rates over the wireless channel is increasing, research on MIMO is antenna systems has gained attention from the academia and the industry.

The information-theoretic aspects of MIMO systems were considered by several authors. In 1987, Winters [41] studied the capacity of MIMO systems and provided results that demonstrated the capacity gains of using multiple antennas at the transmitter and receiver as opposed to the employment of a single antenna at both sides. Then, Telatar [2]; and Foschini and Gans [56] independently derived the capacity of MIMO systems in 1995 and 1998, respectively. They demonstrated that the capacity of a MIMO system increases linearly with the number of transmit antennas when communicating over an independent and identically distributed (i.i.d.) flat Rayleigh channel provided that the number of receive antennas is equal or greater than the number of transmit antennas. In 1994, Seshadri and Winters [13] proposed a transmitter diversity signalling scheme. In 1998, Tarokh et al. [42] improved the diversity scheme of [13] by removing the restriction imposed by the delay element in the transmitter and employing trellis codes as opposed to the simple repetition code used in [13]. This class of codes were referred to in [42] as Space-Time Codes, which include the delay diversity scheme of Seshadri and Winters [13] as a special case. It was further shown in [42] that the proposed space-time trellis codes (STTCs) were capable of providing a diversity order equal to the number of transmit antennas as well as a coding gain that is proportional to the complexity of the code, which is measured based on the number of states in the trellis. These codes, however, are limited by their decoding complexity, which increases exponentially as a function of the diversity level and the transmission rate. In 1998, about seven months after introducing the space-time codes [42], Alamouti [3] discovered a marvellous transmit diversity scheme using two transmit antennas. A key advantage of Alamouti's scheme is the simple linear processing at the receiver, which is based on maximum-likelihood (ML) detection. The decoding algorithm proposed in [3] can be generalized to an arbitrary number of receivers. Alamouti's achievement inspired Tarokh et al. [43, 44] to generalise the transmit diversity scheme to an arbitrary number of transmit antennas, initiating the concept of space-time block codes (STBCs) in 1999. The STBCs provide the same diversity gain as the STTCs, with lower decoding complexity, when employing the same number of transmit antennas. However, a disadvantage of these STBCs when compared to the STTCs is that they provide no coding gain.

STBCs and STTCs are examples of MIMO diversity maximization schemes in which the signals transmitted from different antennas are jointly designed for the sake of minimizing the error rate. There is another class of MIMO schemes that focuses on maximizing the data rate through spatial multiplexing where signals are transmitted independently from each antenna. An example of this class is the scheme proposed by Foschini [56], which was called vertical Bell Laboratories layered space-time (V-BLAST) scheme. The optimum detection method for this scheme is ML detection where all signal space, which consists of all possible combinations of modulation symbols, is exhaustively searched in order to find the combination that minimizes the distance metric when compared to the received signal. However, the size of the signal space becomes huge for large number of transmit antennas or high-order modulations rendering exhaustive search impractical if not impossible. However, detection methods based on zero-forcing (ZF) techniques, which include only matrix inversion to find the best estimate, could be employed. ZF techniques are simple but less accurate when compared to ML detection. A detection method that provides a better compromise between complexity and accuracy was proposed in [33] and is known as nulling and cancelling with optimum ordering. A recently proposed detection method that has attracted considerable attention is sphere decoding [68], which provides an accurate estimation similar to the optimum ML detection but with far less complexity.

1.3.3 Coded Modulation

In 1974, Massey [35] suggested the joint design of the channel encoder and the modulator as a single entity, which is later called a coded modulation system. By doing this, the loss due to the expansion of the signal constellation can be overcome and a significant coding gain can be obtained by using relatively simple codes. Coded modulation is now a popular and powerful technique to improve the performance of digital communication systems whose bandwidth is limited.

In 1982, Ungerboeck introduced Trellis Coded Modulation (TCM) system as a bandwidth efficient signaling over an Additive White Gaussian Noise (AWGN) channel [36]. He mentioned that the primary parameter which determines the code's performance is the minimum free Euclidean distance, and not the minimum Hamming distance. This goal can be achieved by carefully design the signal labeling. In order to get the best labeling, he proposed the signal set expansion, which is the partitioning of signal constellation into subset with increasing intra-subset Euclidean distance. By doing this, TCM achieves large coding gain without reducing bandwidth efficiency. In an example by Ungerboeck [37], TCM with a four-state, rate-2/3 code, and 8-PSK modulation performs 3 dB better than uncoded QPSK, while maintaining the same spectral efficiency.

As mentioned above, the design criterion for TCM is to maximize the minimum intra-subset Euclidean distance [36]. When TCM is designed for fully interleaved Rayleigh fading channels, the design criterion shifts to providing a large diversity order of the coded modulation system [38]. Since Ungerboeck TCM was originally designed for AWGN channels, it usually yields low diversity order in

Rayleigh fading channels. Therefore, TCM performance degrades over Rayleigh fading channels.

An alternative approach to improve TCM was proposed by Zehavi [20] in 1992, which increases the diversity order of TCM. The problem of Ungerboeck TCM that was recognized by him was the fact that it is based on a symbol-by-symbol interleaver in conjuction with a trellis code, so the order of diversity is limited to the minimum number of distinct symbols along any error event. Zehavi then proposed a coded system, which is later called Bit-Interleaved Coded Modulation (BICM), based on bit interleaver. This results in the increasing of diversity order to the minimum number of distinct bits along any error event. Also, unlike TCM, BICM treats coding and modulation as two separate entities. From practical point of view, this allows more flexibility in the design and implementation, compared with the joint code-modulation approach by TCM.

As a bandwidth efficient coded modulation scheme, BICM [20, 21] has been shown to be able to achieve large coding gain over fading channels, and thus has been widely accepted to current wireless local area networks (WLAN) standards, e.g., IEEE 802.11, and the European HiperLAN/2 [40]. However, there is a drawback of BICM due to the increase of time diversity, namely the decrease of minimum free Euclidean distance, and this leads to a performance degradation over AWGN channels when compared to Ungerboeck TCM [20, 21].

Since the invention of turbo codes [39], iterative decoding has also been applied to BICM. It is shown in [22][23][25][26] that with iterative decoding, BICM shows an excellent error performance not only in the fading channels, but also in AWGN channel. More specifically, [23] shows that when signal labeling is carefully designed, iterative decoding can increase the free Euclidean distance, while at the same time retaining the large Hamming distance. This makes BICM with iterative decoding (BICM-ID) greatly outperform TCM and compares favorably with bandwidth-efficient turbo TCM [39, 23].

Among several distinct approaches to construct space-time codes, e.g. Bell Laboratories Layered Space-Time (BLAST) architecture and the space-time trellis code, Bit-Interleaved Space-Time Coded Modulation (BISTCM-ID), which combines serial concatenation of BICM-ID with Space-Time Block Code (STBC), can effectively capture the diversity to provide robust performance under wide variety of fading conditions [57].

1.4 Thesis Outline and Novel Contributions

The thesis is concerned with the problem of reliable wireless transmission over a fast Rayleigh fading environment. We consider the single user transmission with the particular cases of point to point transmission using BICM-ID and two-points to point transmission using BI-STCM-ID. In addition to this we consider transmission over frequency selective channel and the design of space time frequency signaling with full diversity. Finally, the multiuser transmission is considered from an information theoretic point of view. The main contributions is this thesis can be summarized as follows:

- A new multidimensional constellation labeling for BICM-ID based on Reactive Tabu Search (RTS) is designed for 8-PSK which outperforms two dimensional schemes in the whole SNR region. The scheme combines the multidimensional labeling with a designed interleaver.

  Aeman S. Mohammed, Yongxiang Gong and Martin Bossert, "On Multidimensional BICM-ID with 8-PSK Constellation Labeling" *Proc. of IEEE PIMRC 2007*, Sept. 2007, Athens, Greece.

- A BI-STCM-ID with multidimensional labeling is proposed. This signaling scheme results in a high increase in performance without additional complexity compared to the transmission without joint labeling of the space time signals. This work is published in

1 Introduction & Motivation

Aeman S. Mohammed, Wahyu Hidayat and Martin Bossert, "Multidimensional 16-QAM Constellation Labeling of BI-STCM-ID with the Alamouti Scheme" *Proc. of IEEE WCNC 2006*, April 2006, Las Vegas, USA.

- A simple $2 \times 2 \times 2$ Space Time frequency transmission scheme for OFDM transmission with full-rate and full-diversity is constructed. This scheme results in low decoding complexity thank the real rotation matrix. The scheme benefits from the additional frequency diversity and outperforms Alamouti scheme with more than 8 dB. This scheme is presented in

    Aeman S. Mohammed and M. Bossert. "A simple 2x2x2 full-rate full diversity space-time-frequency transmission scheme". *In Proc. 9th International OFDM Workshop*, pages 275-278, Dresden, Germany, September 2004.

- We noted that a broadcasting approach to the multiuser downlink transmission could be more efficient than the classical multiple access approach used in the current wireless transmission standards. Based on the concept of Information Combining (IC) and simulations of simple two user transmission we illustrated the fundamental advantage of broadcast transmission. We presented this idea on

    Martin Bossert and **Aeman S. Mohammed**. Downlink transmission as broadcast channel. *In Proc. International Symposium on Information Theory and its Applications*, pages 116-119, Parma, Italy, October 2004.

Chapter 2

Iterative Decoding, Turbo Codes and Orthogonal Space-Time Block Codes

In this chapter we explain the fundamentals of the concepts and the definitions used elsewhere in the thesis. We introduce the concepts of iterative decoding and turbo codes. Then we describe the basics of coded modulation schemes. We will also discuss the basic concepts behind some Space Time Block Codes STBCs.

2.1 Iterative Decoding and Turbo Codes

Figure 2.1 shows block diagrams of serial and parallel concatenated codes. Although the figures only show one interleaver for simplicity, it is possible to have more than one interleaver. The information symbols **u** are drawn from the set 0,1 with equal probabilities. In the serially concatenated scheme, the superscripts o and i refer to the outer and inner encoders. Hence, $\mathbf{u}^o$ and $\mathbf{u}^i$ corresponds to the input and $\mathbf{c}^o$ and $\mathbf{c}^i$ corresponds to the output, of outer and inner encoders, respectively. The encoding is different for parallel encoder where the information bit stream **u** is encoded by two encoders, with the input to the second encoder, $\mathbf{u}^2$, being a permuted version of **u**. For both these schemes, the encoded output symbols are fed into modulators that map the input symbols into streams of continuous time waveforms to transmit over the channel. In the detector, a set of set of sufficient statistics is obtained by first matched-filtering then sampling, the received waveform. this set of observables becomes the input to the iterative decoder [50].

Figure 2.2 shows an iterative decoder for serially concatenated codes. The block named deinterleaver does exactly the reverse action of the interleaver.

The only difference between a decoder of parallel concatenated codes with the one in Figure 2.2, is he manner in which the different modules in the iterative decoder are connected to one another [50]. The input output relationships of the modules are the same for both serial and parallel decoder realization. According to the figure, we see that the Soft-Input-Soft-Output (SISO) module is the core element in an iterative decoder since it carries out the decoding of the constituent encoders in the transmitter.

To distinguish the input and the output of a certain module, we use the letters I and O. The notation used in the figure can be explained easily using an example. The values $P(\mathbf{c}^i, I)$ denote the input conditional PDFs for random vector $C^i = (C_1^i, C_2^i, ..., C_N^i)$ given the observation vector $R^i = (R_1^i, R_2^i, ..., R_N^i)$. Note that $c^i = (c_1^i, c_2^i, ..., c_N^i)$ and $r^i = (r_1^i, r_2^i, ..., r_N^i)$ are outcomes of C^i and R, respectively.

Figure 2.3 shows the SISO module viewed as a four-port device. It can be said that the SISO module

9

2 Iterative Decoding, Turbo Codes and Orthogonal Space-Time Block Codes

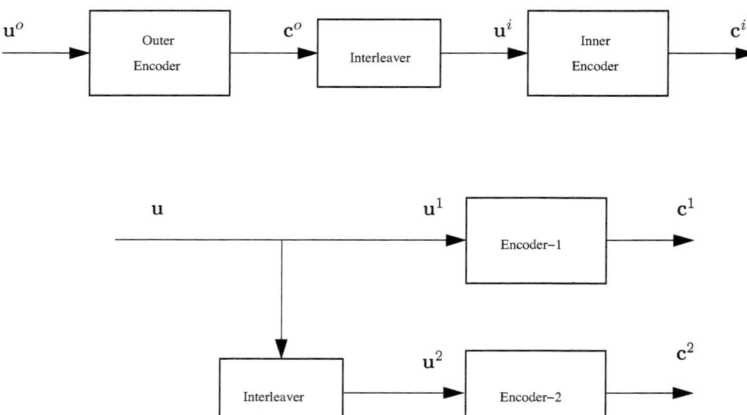

Figure 2.1: Serial and parallel concatenated coding schemes

accepts a priori probability distributions $P(\mathbf{u}, I)$, $P(\mathbf{c}, I)$ at its input, and outputs a posteriori probability distributions $P(\mathbf{u}, O), P(\mathbf{c}, O)$ based on inputs and code constraints. The set of operations performed in the SISO module is often referred to as the SISO algorithm. It should be emphasized that the SISO algorithm is a generalization of the classical BCJR algorithm [51], which carries out symbol-by-symbol MAP detection. The only difference between the two is that the former is capable of handling parallel edges in its trellis diagram. Apart from that, both are the same. Hence, one could say that BCJR is the core decoding algorithm used in turbo (iterative) decoders.

We assume that each input symbol $\mathbf{u}_k$ consists of k_0 bits $u_k^j, j = 1, ..., k_0$ where $u_k^j \in 0, 1$ and each coded symbol $(c)_k$ consists of n_0 bits $c_k^j, j = 1, ..., n_0$ where $c_k^j \in 0, 1$. With these definitions, for a finite time index set, k=1,....,N, the SISO algorithm calculates the output probability distributions for $P(\mathbf{u}_k^j, O), P(\mathbf{c}_k^j, O)$ at time k through the following expressions [49]

$$P(\mathbf{c}_k^j, O) = H_c \cdot \sum_{e:c_k^j} A_{k-1}[s^S(e)] P[u_k(e); I]$$
$$\times (\prod_{i=1, i \neq j}^{n_o} P[c_k^i(e); I]) B_k[s^E(e)], \quad (2.1)$$

$$P(\mathbf{u}_k^j, O) = H_u \cdot \sum_{e:c_k^j} A_{k-1}[s^S(e)] P[u_k(e); I] (\prod_{i=1, i \neq j}^{k_o} P[u_k^i(e); I])$$
$$\times P[c_k(e), I] B_k[s^E(e)]. \quad (2.2)$$

Here, H_c, H_u are normalization constants such that the probabilities in 2.1 and 2.2 add to one when all possible values are considered. Note that the summation is taken over all the edges (e) in the trellis section subjected to the constraints in the summation condition. We will only mention that each edge

2.1 Iterative Decoding and Turbo Codes

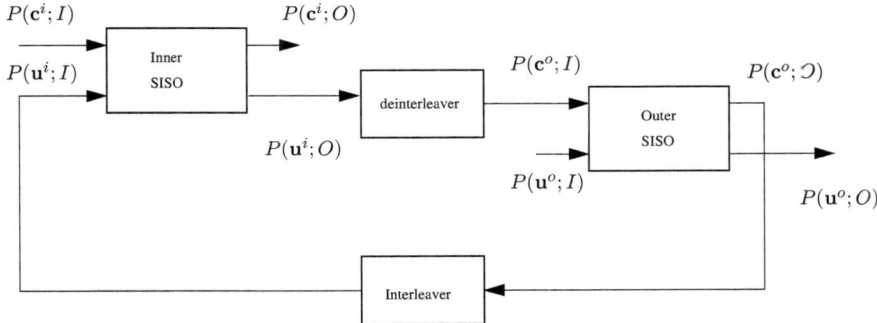

Figure 2.2: Iterative decoder for serially concatenated coding scheme

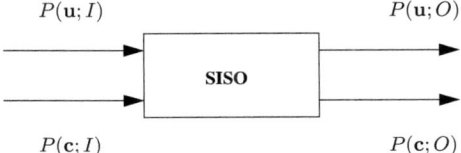

Figure 2.3: Input output relationship of the SISO module

in the trellis diagram is uniquely identified by its starting state (s^S), ending state (s^E), corresponding information symbol and code symbol. See [49] for further details.

The values of $A_k[.], B_k[.]$ for each time index are obtained using forward and backward recursions that satisfy following recursive relationships [49],

$$A_k(s) = \sum_{e:s^E(e)=s} A_{k-1}[s^S(e)] P[u_k(e); I] \times P[c_k(e); I], \qquad (2.3)$$

$$B_k(s) = \sum_{e:s^E(e)=s} B_{k+1}[s^S(e)] P[u_{k+1}(e); I] \times P[c_{k+1}(e); I]. \qquad (2.4)$$

The initial values for these recursions are given by

$$A_0 = \begin{cases} 1, & s = S_0 \\ 0, & \text{otherwise,} \end{cases} \qquad (2.5)$$

$$B_N = \begin{cases} 1, & s = S_N \\ 0, & \text{otherwise.} \end{cases} \qquad (2.6)$$

2 Iterative Decoding, Turbo Codes and Orthogonal Space-Time Block Codes

We have assumed that S_0 and S_N are the initial and final states of the encoder. It should be noted that the output probability distributions of the SISO, $P(u_k^j, O)$ and $P(c_k^j, O)$, are computed based on the code constraints and are obtained using the probability distributions of all bits of the sequence, except the probability distributions of the jth bit within the kth symbol. The probabilities are referred to as the *extrinsic* information in turbo code literature. They are views as the added values by the SISO module to the a priori probability distributions $P(u_k^j, O)$ and $P(c_k^j, O)$, disregarding the contribution by themselves [49].

The operation of the iterative decoder is explained as follows. Initial *a priori* probability $P(\mathbf{u};I)$ in the inner SISO are equal for all symbols since no *a priori* probability information is available before starting the iterative process. In the first iteration, the inner SISO is fed with equal *a priori* probabilities, and the outputs from he receiver front-end. The SISO module calculates the input conditional PDFs $P(\mathbf{c};I)$ from $\mathbf{r}$. These two inputs ($P(\mathbf{u};I)$ and $P(\mathbf{c};I)$) are processed by the inner SISO module to compute the corresponding extrinsic information according to 2.1 and 2.2. The extrinsic information correspond to the information symbols are then passed to the deinterleaver that deinterleaves the extrinsic information at its input before passing the deinterleaved output as the input to the outer SISO. Note that extrinsic information for coded bits is never used, hence the step to calculate these can be omitted.

The outer SISO module also has two inputs, and a similar procedure as above is repeated in the outer SISO. From he outputs of the outer SISO, the extrinsic information corresponding to the coded symbols is then interleaved and fed back to the inner SISO as the *a priori* probabilities for the inner information symbols for the next iteration. The iterations continue until the stopping criterion for the decoder is satisfied. finally, the outer decoder takes hard decisions on the *a posteriori* probabilities of the information symbols, and this is the detected information sequence for the iterative decoder.

2.2 Orthogonal Space-Time Block Codes

Assuming a Generalized Complex Orthogonal Space Time Block Code G_n of size $p \times n$ matrix with symbol entries in the code as $(0, \pm x_1, \pm x_1^*, \pm x_2, \pm x_2^*, \cdots, \pm x_k, \pm x_k^*)$ where $p \neq n$ (non square matrix) for p and n denoting the number of time slots and the number of transmit antennae, respectively. According to the interpretation of Radon-Hurwitz transformation theorem, no full rate complex rectangular (non-square) orthogonal design exists for $n > 2$ transmit antennas. Tarokh et. al classified generalized complex orthogonal designs for $n > 2$ with code rates lower than 1 (non-full rate). The generalized complex rectangular orthogonal codes of size 8×3 and 8×4 achieving a code rate of 0.5 given in [43] as shown below.

Example of generalized complex orthogonal design of size 8×3 achieving a code rate of 0.5

$$\begin{pmatrix} x_1 & x_2 & x_3 \\ -x_2 & x_1 & -x_4 \\ -x_3 & x_4 & x_1 \\ -x_4 & -x_3 & x_2 \\ x_1^* & x_2^* & x_3^* \\ -x_2^* & x_1^* & -x_4^* \\ -x_3^* & x_4^* & x_1^* \\ -x_4^* & -x_3^* & x_2^* \end{pmatrix}. \tag{2.7}$$

Example of generalized complex orthogonal design of size 8×4 achieving a code rate of 0.5

$$\begin{pmatrix} x_1 & x_2 & x_3 & x_4 \\ -x_2 & x_1 & -x_4 & x_3 \\ -x_3 & x_4 & x_1 & -x_2 \\ -x_4 & -x_3 & x_2 & x_1 \\ x_1^* & x_2^* & x_3^* & x_4^* \\ -x_2^* & x_1^* & -x_4^* & x_3^* \\ -x_3^* & x_4^* & x_1^* & -x_2^* \\ -x_4^* & -x_3^* & x_2^* & x_1^* \end{pmatrix}. \tag{2.8}$$

Siavash Alamouti in [3] proposed a code of size 2×2 utilizing space diversity such that the two symbols are sent from the two transmit antennas simultaneously in the first time slot and the second time slot whereas the conjugates of the same symbols sent again in both the time slots to the only receive antenna with the symbols of the code from a complex constellation only. This scheme proposed by Alamouti is classified as a unique case of Complex Orthogonal design by Tarokh in [43].

2.2.1 Encoding of the Alamouti Scheme

A full rate square Complex Orthogonal Space Time Code (COSTBC) is a matrix of size $p \times n$ with the symbols values of the code as $(\pm x_1, \pm x_2, \cdots, \pm x_n)$ and $(\pm x_1^*, \pm x_2^*, \cdots, \pm x_n^*)$ where p denotes the number of time slots needed to transmit the whole code and n the number of transmit antennas with $p = n$ signifying a complex square orthogonal design. Considering the symbol constellation $\mathcal{A}$ with 2^b constellation points where b corresponds to the number of bits necessary to represent a symbol. The bits nb entering the encoder generated by the bit generator are mapped to complex constellation symbols $(x_1, x_2, \cdots, x_n)$ to arrive at a matrix entries which would be orthogonal pairwise.

But assuming a square COSTBC of size 2×2 matrix, one talks explicitly of Alamouti code with 2 transmit antennas in which complex symbol x_1 is transmitted from antenna I and symbol x_2 from antenna II simultaneously in the first time slot while negative conjugate of the second symbol $-x_2^*$ and conjugate of the first symbol x_1^* are transmitted simultaneously in the second time slot to arrive at the matrix as shown below.

$$G_2 = \begin{pmatrix} x_1 & x_2 \\ -x_2^* & x_1^* \end{pmatrix} \tag{2.9}$$

Alamouti scheme can be applied to an arbitrary complex signal constellation such as PSK and QAM. The added advantage which Alamouti code offers is the full code rate R shown in equation below to be equal to 1.

$$R = k/n \tag{2.10}$$

where k denotes the information symbols sent in a block of p time slots with n transmit antennas. As two symbols are noticed to be sent in a block of $p = 2$ time slots with $n = 2$ transmit antennas, then a code rate of 1 is achieved. The channel model which we will use throughout the thesis for the two transmit antennas case is outlined below

2.2.2 Channel Model

Considering a transmission system with 2 transmit antennae and 1 receive antenna as shown in the Figure 2.4. Quasi-static flat fading is assumed, i.e. the channel path gain remains constant for a complete symbol period. The transmitted symbols in the first and second time slots pass through the Rayleigh fading channel with channel path gains $h_1(t)$ and $h_2(t)$ between transmit antenna I to the receiver and transmit antenna II to the same receiver respectively.

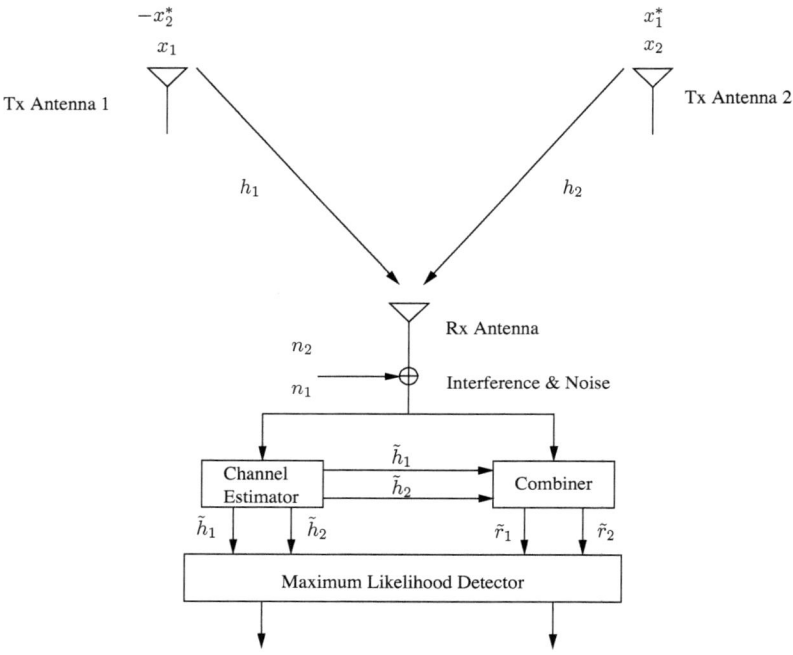

Figure 2.4: Alamouti Scheme Model with 2 Transmit and 1 Receive Antennas

As shown in the Figure 2.4, the transmitted symbols after passing through the channel with the induction of noise and multipath fading effects can be given in the form of a received vector $y(t)$ as shown below.

$$y(t) = \begin{pmatrix} x_1 & x_2 \\ -x_2^* & x_1^* \end{pmatrix} \cdot \begin{pmatrix} h_1(t) \\ h_2(t) \end{pmatrix} + \begin{pmatrix} n_1(t) \\ n_2(t) \end{pmatrix} \qquad (2.11)$$

where n_1 and n_2 corresponds to the noise added in first and second time slots respectively.

The code matrix defined in equation 2.9 satisfies the Radon-Hurwitz Orthogonal Criteria. According to Radon-Hurwitz Theorem, an OSTBC design G_n of size $p \times n$ square matrix has to hold the following equation in order to be orthogonal [43].

$$G_n G_n^T = D_n \qquad (2.12)$$

where D_n is a diagonal matrix with diagonal elements of the form

$$(l_1^i |x_1|^2 + l_2^i |x_2|^2 + \cdots + l_k^i |x_k|^2) \qquad (2.13)$$

2.2 Orthogonal Space-Time Block Codes

with all the coefficients $(l_1^i, l_2^i, \cdots, l_k^i)$ all strictly positive numbers.

Thus, for Alamouti codes, the above equation becomes

$$G_2 G_2^T = \begin{pmatrix} x_1 & x_2 \\ -x_2^* & x_1^* \end{pmatrix} \begin{pmatrix} x_1^* & x_2^* \\ -x_2 & x_1 \end{pmatrix} \quad (2.14)$$

$$G_2 G_2^T = \begin{pmatrix} x_1 x_1^* + x_2 x_2^* & -x_2^* x_1^* + x_1^* x_2^* \\ -x_1 x_2 + x_1 x_2 & x_2^* x_2 + x_1 x_1^* \end{pmatrix} \quad (2.15)$$

$$G_2 G_2^T = \begin{pmatrix} (|x_1|^2 + |x_2|^2) & 0 \\ 0 & (|x_1|^2 + |x_2|^2) \end{pmatrix} \quad (2.16)$$

$$G_2 G_2^T = (|x_1|^2 + |x_2|^2) \begin{pmatrix} 1 & 0 \\ 0 & 1 \end{pmatrix} \quad (2.17)$$

$$G_2 G_2^T = (|x_1|^2 + |x_2|^2) I_2 \quad (2.18)$$

where I_2 is the identity matrix of order 2.

As mentioned before that according to the interpretation of Radon-Hurwitz theorem, the Alamouti scheme is the only valid complex orthogonal symbols scheme with number of transmit antennae equal to 2 ($n = 2$) and no complex orthogonal design of full rate with number of transmit antennae greater than 2 exists [43].

2.2.3 Detection of the Alamouti Scheme

In accordance to the Alamouti scheme shown in figure 2.4 with channel path gains of $h_1(t)$ and $h_2(t)$ assuming the fading coefficients to be constant over two consecutive symbols signifying a quasi-static channel model.

The received signals $r_1(t)$ and $r_2(t+T)$ at time slots t and $(t+T)$ respectively with symbol period T are summed up as:

$$\begin{aligned} r_1 &= r_1(t) = h_1(t) x_1(t) + h_2(t) x_2(t) + n_1(t) \\ r_2 &= r_2(t+T) = -h_1(t+T) x_2^*(t) + h_2(t+T) x_1^*(t) + n_2(t) \end{aligned} \quad (2.19)$$

Using the assumption of quasi-static channel such that

$$h_i(t) = h_i(t+T) = h_i \quad (2.20)$$

for $i = 1, 2, \cdots, n$

Thus the equations simply reduces to

$$\begin{aligned} r_1(t) &= r_1 = h_1 x_1 + h_2 x_2 + n_1 \\ r_2(t+T) &= r_2 = -h_1 x_2^* + h_2 x_1^* + n_2 \end{aligned} \quad (2.21)$$

Assuming that the channel estimator estimates the channel perfectly i.e. the receiver has perfect knowledge of the Channel State Information (CSI). The channel impulse response estimated by the channel estimator at the receiver side is ideally assessed to be exactly the same as the actual channel impulse response encountered by the transmit signals in the medium $h_i = \alpha_i e^{-\phi_i}$. Thus, the combiner gives the signals as:

$$\begin{aligned} \tilde{r}_1 &= h_1^* r_1 + h_2 r_2^* = (\alpha_1^2 + \alpha_2^2) x_1 + h_1^* n_1 + h_2 n_2^* \\ \tilde{r}_2 &= h_2^* r_1 - h_1 r_2^* = (\alpha_1^2 + \alpha_2^2) x_2 - h_1 n_2^* + h_2^* n_1 \end{aligned} \quad (2.22)$$

2 Iterative Decoding, Turbo Codes and Orthogonal Space-Time Block Codes

The signals received after the combiner are sent to Maximum Likelihood (ML) Decoder which decides for the most probable symbol by applying the decision criteria expressed in the following equations: Choosing x_i iff

$$(\alpha_1^2 + \alpha_2^2) \mid x_i \mid^2 + d^2(\tilde{r}_1, x_i) \leq (\alpha_1^2 + \alpha_2^2) \mid x_k \mid^2 + d^2(\tilde{r}_1, x_k) \forall \ i \neq k \qquad (2.23)$$

or for PSK signals choosing x_i iff

$$d^2(\tilde{r}_1, x_i) \leq d^2(\tilde{r}_1, x_k) \qquad \forall \ i \neq k \qquad (2.24)$$

The signal from the two transmit antennae are simply added to form a combined signal as if transmitted from a single transmit antenna. Thus, the decoding of two transmit antennae remain as simple as employing a single transmit single receive antenna however the Alamouti scheme doubles the diversity order of the system to 2 which is given as nm where n corresponds to the number of transmit antenna and m to the receive antennas.

Chapter 3

Bit Interleaved Coded Modulation with Iterative Decoding (BICM-ID)

The idea of bit-interleaving for coded modulation can be dated back to Viterbi's pragmatic approaches to TCM [47], or even earlier. However, it was Zehavi who first realized the potential of bit-interleaving for Rayleigh fading channels [20]. This technique was later known as Bit-Interleaved Coded Modulation. BICM has shown to achieve large coding gain over fading channels [20, 21]. The main advantage of BICM is that the diversity order can be increased to the minimum Hamming distance of the code by bitwise interleaving. Besides the large diversity order, BICM also provides more flexibility in the design of the encoder and modulation individually, which is very attractive for practical use.

Although BICM performs well over fading channels because of an increase in diversity order, one pitfall of BICM is the reduction of free Euclidean distance over Gaussian channel due to the "random modulation" caused by bit-interleaving [20]. Recall that the free Euclidean distance is the key parameter in determining the performance of BICM over AWGN channel, this "random modulation" therefore causes performance degradation of BICM system over AWGN channel.

In [22, 23], an iterative decoding with hard- and soft-decision feedback for BICM was proposed. By careful design of signal labeling, BICM with iterative decoding (BICM-ID), can overcome the drawback of conventional BICM over AWGN channel by increasing the Euclidean distance with the knowledge of other bit values and this results in a further increase of the error performance.

This chapter will review the system of BICM and BICM-ID. First, the system model of BICM and BICM-ID are described. Then, we will review the error bound, design optimization and some parameters that affect the performance of the systems.

3.1 Bit Interleaved Coded Modulation (BICM)

The block diagram of BICM system is shown in Figure 3.1. The transmitter of BICM is a serial concatenation of a convolutional encoder of rate k_c/n_c, a random bit interleaver and a memoryless modulator [21]. The purpose of bit interleaver is to break the sequential fading correlation and to increase the diversity order to the minimum Hamming distance of the convolutional code [20, 21].

Denote the encoder input symbol at time t by $\mathbf{u_t} = [u_t^1, \ldots, u_t^i, \ldots, u_t^{k_c}]$ and the output symbol by $\mathbf{c_t} = [c_t^1, \ldots, c_t^i, \ldots, c_t^{k_c}]$, where u_t^i or c_t^i is the i-th bit in the symbol. The encoder output is bitwise interleaved and each m consecutive bits of the interleaved sequence are grouped to form a channel symbol $\mathbf{v_t} = [v_t^1, \ldots, v_t^i, \ldots, v_t^{k_c}]$. The modulator maps each $\mathbf{v_t}$ to a complex transmitted signal $x_t =$

$\mu(\mathbf{v_t})$ chosen from M-ary constellation χ, where μ is the labeling map and $M = 2^m$.

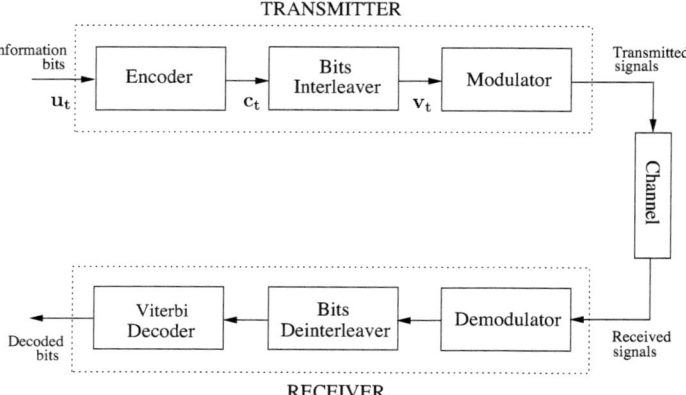

Figure 3.1: The block diagram of a BICM system

In this thesis, we consider a fully interleaved frequency non-selective fading channel and coherent detection. The received signal during each symbol duration can be written as follows :

$$y_t = h_t x_t + n_t \qquad (3.1)$$

where h_t is the Rayleigh random variable representing the fading amplitude of the transmitted signal x_t and n_t is a complex white Gaussian noise with variance $\sigma_I^2 = \sigma_Q^2 = N_0/2$. It is assumed that the cannel fades slowly so that the fading amplitude is constant over one symbol duration. It is also assumed that h_t is perfectly estimated at the receiver. Note that in the case of an AWGN channel, $h_t = 1$.

The receiver of a BICM system includes three elements : the demodulator, the bit-deinterleaver and the convolutional decoder. Due to the presence of random bit interleaver, the true maximum likelihood decoding of BICM requires joint demodulation and convolutional decoding. Therefore it is too complicated to be implemented in practice. As a trade-off between the complexity and the error performance, in [20] Zehavi suggested a suboptimal decoding method that includes two separate steps : the bit metric computation, which is done in demodulator, and the decoding of convolutional code by the Viterbi algorithm [34].

For each received signal y_t, log-likelihood bit metrics for m bits are computed using ML rules as follow :

$$\lambda(v_t^i = b) = \log p(y_t | v_t^i = b, h_t)$$

$$\sim \log \sum_{x_t \in \chi_b^i} p(y_t | x_t, h_t)$$

$$i = 1, 2, \ldots, m; b = 0, 1 \qquad (3.2)$$

where χ_b^i is the subset of χ whose labels have the binary value of b at the i-th bit position. For M-ary constellation, there are $2m$ subsets, each of them has size of 2^{m-1}. The function $p(y_t | x_t)$ is the

3.1 Bit Interleaved Coded Modulation (BICM)

probability density function (pdf) of a received signal y_t given the signal x_t was transmitted. The notation $\sim$ in (3.2) indicates replacement by an equivalent statistic.

With two-dimensional M-ary signal constellation χ, $p(y_t|x_t)$ is given as :

$$p(y_t|x_t) = \frac{1}{2\pi\sigma^2} \exp\left[-\frac{(y_{t1} - h_t x_{t1})^2 + (y_{t2} - h_t x_{t2})^2}{2\sigma^2}\right], \quad (3.3)$$

where (y_{t1}, y_{t2}) and (x_{t1}, x_{t2}) are the components of y_t and x_t in Euclidean space, respectively.

For practical implementation, the log sum calculation in (3.2) is approximated resulting in a suboptimal maximum log-likelihood bit metric written as :

$$\lambda(v_t^i = b) \approx \max_{x_t \in \chi_b^i} \log p(y_t|x_t, h_t) = -\min_{x_t \in \chi_b^i} \| y_t - h_t x_t \|^2. \quad (3.4)$$

To make it clearer, an example on how to calculate the bit metrics for BICM system using 8-PSK constellation is given in Figure 3.2.

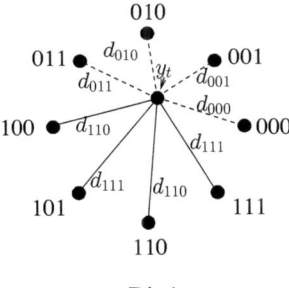

Bit 1

Figure 3.2: Bit metric calculation for BICM

The bit metric for the first bit with a value of 1 is :

$$\lambda(v_t^1 = 1) = \min(d_{100}^2, d_{101}^2, d_{110}^2, d_{111}^2) \quad (3.5)$$

and the bit metric for the first bit with a value of 0 is :

$$\lambda(v_t^1 = 0) = \min(d_{000}^2, d_{001}^2, d_{010}^2, d_{011}^2) \quad (3.6)$$

where $d_{x_t}^2 (x_t = 000, \ldots, 111)$ is given by $\| y_t - h_t x_t \|^2$.

From equation (3.4) and Figure 3.2, it is obvious that each bit metric is computed based on the minimum squared Euclidean distance between the received signal y_t and the signal point x_t over the subset χ_b^i.

Finally, the branch metrics are calculated using the formula :

$$L(v_t^i) = \lambda(v_t^i = 1) - \lambda(v_t^i = 0) \quad (3.7)$$

to be fed to the input of Viterbi decoder.

3 Bit Interleaved Coded Modulation with Iterative Decoding (BICM-ID)

The union bound of the BER for a BICM system employing a convolutional code with rate k_c/n_c, a constellation χ and a labeling μ is given by [21]:

$$P_b \leq \frac{1}{k_c} \sum_{d=d_f}^{\infty} W_I f(d,\chi,\mu), \qquad (3.8)$$

where d_f is the minimum free Hamming distance of the code and W_I is the total input weight of error events at Hamming distance d. The function $f(d,\chi,\mu)$ denotes the Pairwise Error Probability (PEP) of BICM system, the probability that a code sequence $\mathbf{c}$ is transmitted but a code sequence $\hat{\mathbf{c}}$ is selected at the decoder. This function depends only on the Hamming distance d, signal constellation χ and the constellation mapping μ, and can be written in the form of [21] :

$$f(d,\chi,\mu) \leq \frac{1}{2\pi j} \int_{\alpha-j\infty}^{\alpha+j\infty} [\psi_{ub}(s)]^d \frac{ds}{s}, \qquad (3.9)$$

where

$$\psi_{ub}(s) = \frac{1}{m2^m} \sum_{i=1}^{m} \sum_{b=0}^{1} \sum_{x \in \chi_b^i} \sum_{z \in \chi_{\bar{b}}^i} \Phi_{\Delta(x,z)}(s),$$

$\bar{b}$ is the complement of b, α belongs to the intersection of the regions of convergence of $\Phi_{\Delta(x,z)}(s)$ with the real positive line, and $\Phi_{\Delta(x,z)}(s)$ is the Laplace transform of the pdf of the metric difference $\Delta(x,z)$. For Rayleigh fading channel, $\Phi_{\Delta(x,z)}(s)$ is given as [26] : ,

$$\Phi_{\Delta(x,z)}(s) = \frac{1}{1 + s(1 - sN_0) \parallel x - z \parallel^2}. \qquad (3.10)$$

When Gray labeling is used, irrelevant error events can be expurgated [21] from (3.9), and this results in BICM Expurgated (BICM EX) bound. The PEP for BICM EX bound can be written as [26] :

$$f(d,\chi,\mu) \leq \frac{1}{2\pi j} \int_{\alpha-j\infty}^{\alpha+j\infty} [\psi_{ex}(s)]^d \frac{ds}{s} \qquad (3.11)$$

where

$$\psi_{ex}(s) = \frac{1}{m2^m} \sum_{i=1}^{m} \sum_{b=0}^{1} \sum_{x \in \chi_b^i} \Phi_{\Delta(x,\hat{z})}(s)$$

and $\hat{z} = \hat{z}(x) \in \chi_{\bar{b}}^i$ denotes the nearest neighbor of x.

Then the PEP of the error floor of BICM system can be numerically evaluated by the Gauss-Chebyschev quadrature method [53] before the bit error probability is calculated using (3.8).

Note that BICM EX bound is formally similar to the BICM union bound, but it includes only one error event ($x \rightarrow \hat{z}$) for each possible transmitted symbol x, rather than all the possible error events for all $z \in \chi_{\bar{b}}^i$, since the number of nearest neighbor of x in only one, that is the symbol $\hat{z} \in \chi_{\bar{b}}^i$ whose bit value is different from x only in i-th position.

3.2 Bit Interleaved Coded Modulation with Iterative Decoding (BICM-ID)

Convolutional encoding introduces redundancy and memory into the signal sequence x_t, which is used as *a priori* information. Yet, the equally likely assumption for $P(x)$ in (3.2) fails to use this information,

3.2 Bit Interleaved Coded Modulation with Iterative Decoding (BICM-ID)

primarily because it is difficult to specify in advance of any decoding. However, the *a priori* information of the transmitted signal is reflected in the decoding results. Therefore, the use of iterative decoding can be used as *a priori* information, so that the feedback from the strong data section (which is less affected by the channel noise) can remove the ambiguity in the high-order demodulation and enhance the decoding of weak data section (which is the subject to undesirable noise patterns).

Recall that in the conventional BICM system, the bit metric for each coded bit is computed at the demodulator as in (3.4) based on ML rule. This metric is essentially a simplification of the *a posteriori* probabilities in the Maximum *A posteriori* Probability (MAP) criterion [48], under the assumption that the transmitted signals are equally likely. The MAP criterion can be calculated as follow :

$$\lambda(v_t^i = b) = \log \sum_{x_t \in \chi_b^i} P(x_t|y_t, h_t). \tag{3.12}$$

Using Bayes rule [48], the *a posteriori* probability $P(x_t|y_t, h_t)$ can be calculated as :

$$P(x_t|y_t, h_t) = \frac{p(y_t|x_t, h_t)P(x_t)}{p(y_t|h_t)}, \tag{3.13}$$

where $P(x_t)$ is the *a priori* probability that the signal x_t is transmitted. The denominator of (3.13) can be expressed as :

$$p(y_t|h_t) = \sum_{i=1}^{M} p(y_t|x_t, h_t) P(x_t). \tag{3.14}$$

From (3.13) and (3.14), it can be observed that the calculation of the *a posteriori* probability $P(x_t|y_t, h_t)$ requires the knowledge of the *a priori* probability $P(x_t)$ and the conditional pdf $p(y_t|x_t, h_t)$. Hence, the statistic for the bit metric $\lambda(v_t^i = b)$ can be written as :

$$\lambda(v_t^i = b) \sim \log \sum_{x_t \in \chi_b^i} p(y_t|x_t, h_t) P(x_t). \tag{3.15}$$

Figure 3.3 shows the block diagram of BICM-ID receiver using soft-decision feedback, where the Viterbi decoder is replaced by the SISO module described in chapter 2.

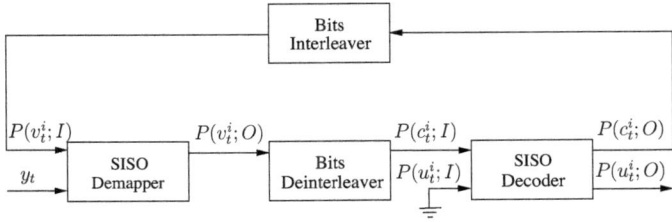

Figure 3.3: The block diagram of BICM-ID receiver with soft-decision feedback

Let c_t denote the deinterleaved version of v_t and u_t denote the input symbol corresponding to c_t. Following notation of [49], $P(q; I)$ denotes the *a priori* probability for a variable q. $P(q; O)$ is the *a posteriori* probability. Note that $P(u_t; I)$ is never available due to the requirement of the information sequence at the receiver.

21

3 Bit Interleaved Coded Modulation with Iterative Decoding (BICM-ID)

On the second round, the extrinsic *a posteriori* probabilities $P(c_t^i; O)$ put out by the SISO decoder are interleaved and fed back as the *a priori* probabilities $P(v_t^i; I)$ to the demodulator. Therefore, *a priori* probabilities for (3.15) can be calculated as [26] :

$$P(x_t) = P(\mu(v_t; I))$$
$$= \prod_{i=1}^{m} P(v_t^i = \tilde{v}_t^i(x_t); I), \qquad (3.16)$$

where $\tilde{v}_t^i(x_t)$ is the value of the i-th bit of the label corresponding to $x_t = \mu(\tilde{v}_t)$. Using (3.15) and (3.16), the extrinsic *a priori* bit probabilities for the second round demodulation can be written as :

$$P(v_t^i = b; O) = P(v_t^i = b|y_t)/P(v_t^i = b; I)$$
$$= \sum_{x_t \in \chi_b^i} \left(P(y_t|x_t) \prod_{j=1, j \neq i}^{m} P(v_t^j = \tilde{v}_t^j(x_t); I) \right). \qquad (3.17)$$

Equation (3.17) shows that we need only the *a priori* probabilities $P(v_t^j; I)$ of the other bits ($j \neq i$) on the same channel symbol v_t when recalculating the bit metrics. The receiver then uses (3.17) to regenerate the bit metrics and iterate demodulation and decoding. After the last round, the final decoded outputs are the hard-decisions based on the extrinsic bit probabilities $P(u_t^i; O)$. This is the total *a posteriori* because $P(u_t^i; I)$ is unused. To reduce the computational complexity, the SISO decoder uses an additive log-MAP algorithm [49].

3.2.1 The Error Bound of BICM-ID System

Due to large coding gain produced by iterative processing, one is most interested in the asymptotic performance of BICM-ID performance to which the iterative decoding converges. Such an asymptotic performance of BICM-ID can be analyzed using Error-Free (EF) bound introduced in [25, 26]. The EF Bounds in [25, 26], both for Rayleigh fading channels and AWGN channel, are basically the modification of the error bounds of BICM system by assuming that error-free feedback is used.

However, the EF bound for the case of AWGN channel derived in [25, 26], which based on the free squared Euclidean distance conditioned on error free feedback (FEDC), sometimes fails to discriminate the differences in the error performance of different mappings [15]. Therefore, a new method to calculate the upper bound of BICM-ID system was derived in [15]. This derivation is still based on BICM union bound (3.8), but using the techniques presented in [54, 55] in deriving the PEP of the systems.

Over Rayleigh fading channel, the PEP of the system can be approximated by [15] :

$$f(d, \chi, \mu) \approx \frac{1}{2} \left[4N_0 \tilde{\delta}_1(\chi, \mu) \right]^d, \qquad (3.18)$$

where

$$\tilde{\delta}_1(\chi, \mu) = \left(\frac{1}{m 2^m} \sum_{i=1}^{m} \sum_{b=0}^{1} \sum_{x \in \chi_b^i} \frac{1}{\| x - \tilde{z} \|^2} \right). \qquad (3.19)$$

Note that $\tilde{\delta}_1(\chi, \mu)^{-1}$ is exactly the harmonic mean Euclidean distance with ideal feedback defined in [25] for BICM-ID system over Rayleigh fading channel.

For the case of AWGN channel, the PEP of the system can be written as [15] :

$$f(d, \chi, \mu) \approx \frac{1}{2} [\delta_2(\chi, \mu)]^d, \qquad (3.20)$$

3.2 Bit Interleaved Coded Modulation with Iterative Decoding (BICM-ID)

where

$$\delta_2(\chi,\mu) = \frac{1}{m2^m} \sum_{i=1}^{m} \sum_{b=0}^{1} \sum_{x \in \chi_b^i} \exp\left(-\frac{\|x-\tilde{z}\|^2}{4N_0}\right). \quad (3.21)$$

From the equation (3.19) and (3.21), it is clear that the signal labeling has a strong influence to the asymptotic performance of BICM-ID system over these two channels. Obviously, the best mappings with respect to the asymptotic performance of BICM-ID system over both Rayleigh fading channels and AWGN channel are the ones that have the smallest values of $\tilde{\delta}_1(\chi,\mu)$ and $\delta_2(\chi,\mu)$.

The asymptotic performance of the error floor of BICM-ID over Rayleigh fading at high SNR can also be approximated as [26] :

$$\log_{10} P_b \simeq \frac{-d_f}{10} \left[(R\tilde{d}_h^2)_{dB} + \left(\frac{E_b}{N_0}\right)_{dB} \right] + const, \quad (3.22)$$

where

$$\tilde{d}_h^2(\mu) = \tilde{\delta}_1(\chi,\mu)^{-1}. \quad (3.23)$$

It can be observed that the code and the labeling have independent impacts on the asymptotic performance, namely, the code controls the slope of the error bound through its minimum free Hamming distance d_f, while the labeling determines the horizontal offset of the error bound through its harmonic mean distance $\tilde{d}_h^2$.

3.2.2 Labeling for Different Modulation Formats

As already discussed in Subsection 3.2.1, signal labeling has a critical influence to the performance of BICM and BICM-ID systems.

In case signal transmission over Rayleigh fading channels, that influence can be quantified by the harmonic mean Euclidean distances. The distance d_h^2 affects the asymptotic performance of the BICM while the distance $\tilde{d}_h^2$ dominates the asymptotic performance of BICM-ID.

For 8-PSK constellation, the comparison among different labeling methods are presented graphically in Figures 3.4 and 3.5, while the numerical results from calculating the harmonic mean of the minimum Euclidean distance before feedback d_h^2 and after feedback $\tilde{d}_h^2$ are shown in Table 3.1 [25].

For 16-QAM constellation, the graphical comparison are shown in the Figure 3.6 and 3.7, and the numerical calculation of d_h^2 and $\tilde{d}_h^2$ are shown in Table 3.2 [26].

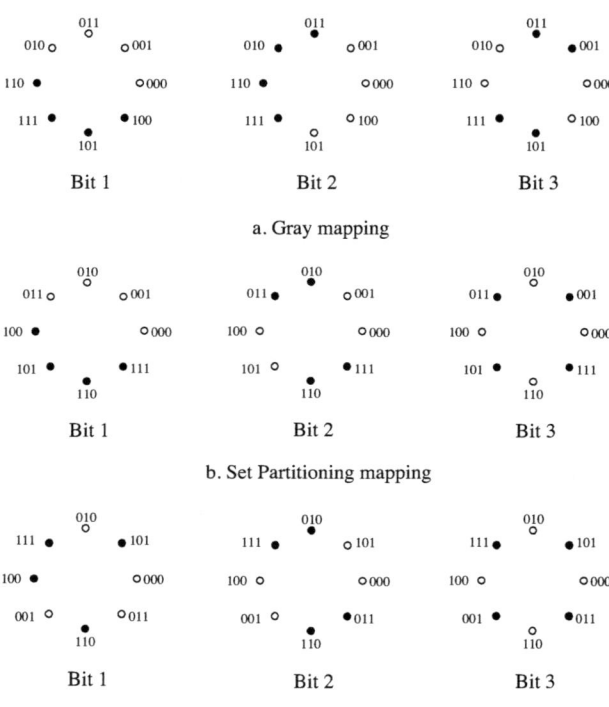

Figure 3.4: Subset partitions of 8-PSK for three different labeling schemes

3.2 Bit Interleaved Coded Modulation with Iterative Decoding (BICM-ID)

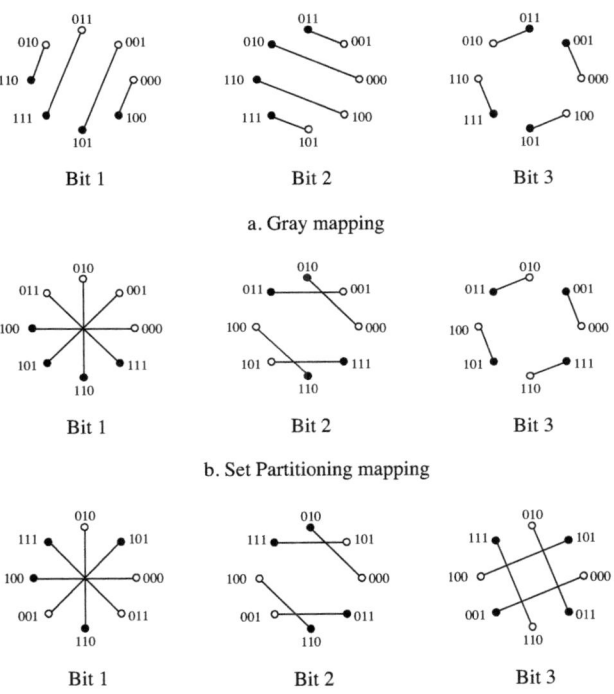

Figure 3.5: 8-PSK channel is converted into four binary channels, each selected by the other two ideal feedback bits

3 Bit Interleaved Coded Modulation with Iterative Decoding (BICM-ID)

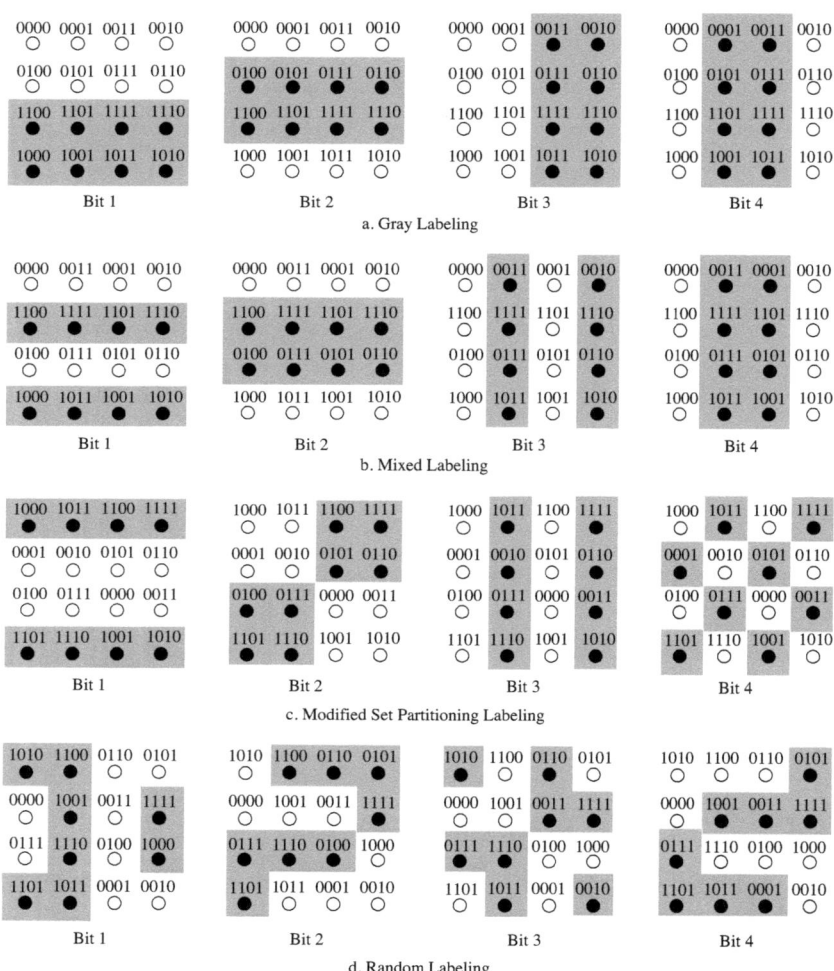

Figure 3.6: Subset partitions of 16-QAM for four different labeling schemes

3.2 Bit Interleaved Coded Modulation with Iterative Decoding (BICM-ID)

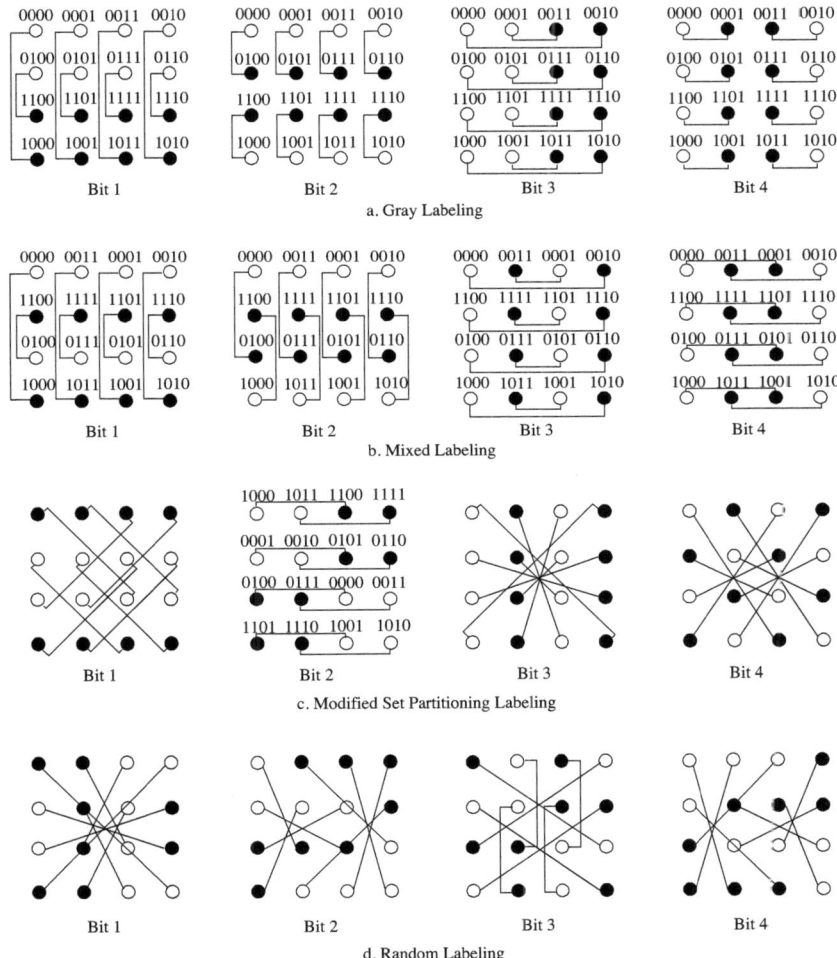

Figure 3.7: 16-QAM channel is converted into eight binary channels, each selected by the other three ideal feedback bits

3 Bit Interleaved Coded Modulation with Iterative Decoding (BICM-ID)

Table 3.1: Comparison of the harmonic mean of the minimum squared Euclidean distance before (d_h^2) and after ideal feedback ($\tilde{d}_h^2$), and the offset gain over Gray labeling for 8-PSK constellation

Labeling	d_h^2 (before)	$\tilde{d}_h^2$ (after)	Offset Gain (dB)
Gray	0.7664	0.8093	0.24
SP	0.6640	1.2209	2.02
SSP	0.5858	2.8766	5.74

Table 3.2: Comparison of the harmonic mean of the minimum squared Euclidean distance before (d_h^2) and after ideal feedback ($\tilde{d}_h^2$), and the offset gain over Gray labeling for 16-QAM constellation

Labeling	d_h^2 (before)	$\tilde{d}_h^2$ (after)	Offset Gain (dB)
Gray	0.7664	0.8093	0.24
SP	0.6640	1.2209	2.02
SSP	0.5858	2.8766	5.74
Random	0.413	2.602	7.23

Figure 3.4 and 3.6 illustrate the subset partitioning for each of m bit positions for different labeling methods. The shaded regions (only shown inside the unit square) correspond to the decision regions for each bits in χ_1^i, while the unshaded one to χ_0^i. These are also the decision regions for each bit if hard-decision detection were made for each bit individually before convolutional decoding. It is obvious that all labeling methods for 8-PSK and those for 16-QAM have the same minimum Euclidean distance between subsets of χ_1^i and χ_0^i, respectively, but a different number of nearest neighbors. Therefore, for conventional BICM, Gray labeling has been considered to be optimal [20, 21] due to the smallest number of nearest neighbors.

Given ideal feedback for all other bits, an 8-PSK as well as 16-QAM constellation is translated to a binary channel selected from four and eight possible pairs, respectively. Figure 3.5 and 3.7 illustrate the increase in the minimum Euclidean distance between subsets. Gray labeling is not the preferred choice for both constellations because most of the binary signal sets resulting from ideal feedback have the same intersignal Euclidean distance as original constellation. These pictures also show that iterative decoding of BICM not only increases the intersignal Euclidean distance, but also reduces the number of nearest neighbors to one, as in the case of Gray labeling for BICM system. This leads to significant improvement over both AWGN and fading channels.

From the numerical calculation of d_h^2 and $\tilde{d}_h^2$ presented in Table 3.1 and 3.2, the offset gains can be calculated. Offset gain is actually the difference in $\tilde{d}_h^2$ (dB) of the BICM-ID system with a particular labeling and d_h^2 (dB) of conventional BICM with Gray labeling. This gives a quick comparison between various labeling schemes with iterative decoding and conventional BICM. In addition, optimization of d_h^2 is done separately from decoding algorithm. Therefore, the offset gain is the asymptotic performance improvement regardless of the code structure used.

It is clear from the discussion above that a labeling that maximizes $\tilde{d}_h^2$ while still having a sufficient large d_h^2 is preferable, so that the system can work well in the first round (conventional BICM), and the feedback decoder can reach its ideal performance after a few iterations.

Now, we will consider the impact of signal labeling on the offset gain. For BICM-ID with the same

3.2 Bit Interleaved Coded Modulation with Iterative Decoding (BICM-ID)

convolutional code, Table 3.1 and 3.2 show that Gray labeling yields the best performance in the first round due to the largest d_h^2. However, the performance gain with feedback is very small. Therefore, Gray labeling is not preferred for BICM-ID system.

For 8-PSK constellation, it is shown in Table 3.1 that SSP labeling has the worst d_h^2, but the largest $\tilde{d}_h^2$ after feedback. Therefore, SSP labeling has the worst performance in the first-round but has the largest asymptotic offset gain. For a 16-QAM constellation, random labeling gives the largest $\tilde{d}_h^2$ and thus the asymptotic offset gain at the cost of having the poorest first round performance. MSP labeling shows good compromise between the first-round performance and the asymptotic iterative decoding performance.

3.2.3 Labeling of BICM-ID as Quadratic Assignment Problem

Quadratic Assignment Problem (QAP) is a kind of combinatorial optimization, given a set of n locations, n facilities, and two $n \times n$ matrices: a flow matrix $\mathbf{F} = \{f_{ij}\}$ where f_{ij} is the flow between facilities i and j; a distance matrix $\mathbf{D} = \{d_{ij}\}$ where d_{ij} is the distance between locations i and j. The objective is to find an optimal permutation p of numbers $\{1, 2, \ldots, n\}$ to minimize the total cost:

$$\min_p \sum_{i=1}^{n} \sum_{j=1}^{n} f_{ij} d_{p(i)p(j)} \tag{3.24}$$

where $p(i), p(j)$ are the locations to which facilities i and j are assigned, respectively.

Considering equation (3.19), maximization of the harmonic mean Euclidean distance over signal mapping μ is equivalent to:

$$\min_\mu \sum_{k=1}^{m} \sum_{b=0}^{1} \sum_{x \in \chi_b^k} \frac{1}{\| x - \tilde{x} \|^2} \tag{3.25}$$

The locations are the signal constellation points $\{x_i | i = 0, \ldots, 2^m - 1\}$. The facilities are the labels $\{0, \ldots, 2^m - 1\}$. And the optimization problem is to find the one-to-one mapping rule $\mu:\{0, \ldots, 2^m - 1\} \to \{x_0, \ldots, x_{2^m-1}\}$, such that the cost function (3.25) is minimized.

The "distance" between two locations x_i and x_j can be defined as :

$$d_{x_i, x_j} = \begin{cases} \frac{1}{\|x_i - x_j\|^2}, & i \neq j \\ 0, & i = j \end{cases} \tag{3.26}$$

and the "flow" between labels i and j is defined as :

$$f_{ij} = \begin{cases} 1, & i \text{ and } j \text{ differ only in one bit position} \\ 0, & \text{otherwise} \end{cases} \tag{3.27}$$

Thus, equation (3.25) becomes

$$\min_\mu \sum_{i=0}^{2^m-1} \sum_{j=0}^{2^m-1} f_{ij} d_{\mu(i), \mu(j)} \tag{3.28}$$

which is a QAP problem as defined in (3.24).

3.2.4 EXIT-chart Analysis

While most studies of concatenated codes with iterative decoding have focused on providing the asymptotic bit error bounds, the convergence properties of iterative decoding schemes have gained a considerable amount of interest. The Extrinsic Information Transfer (EXIT) chart is a visualization method based on bitwise mutual information which provides a tool for studying the convergence behavior of iterative decoding schemes [59]. EXIT chart is important for the main work of this thesis as a tool to analyse the converge behavior of the proposed BITCM-ID system using multidimensional signal labeling.

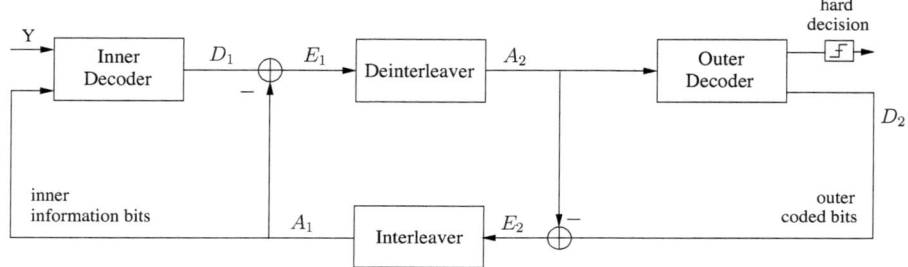

Figure 3.8: Iterative decoding of serially concatenated codes

Figure 3.8 shows the iterative decoding of BICM-ID system. For each iteration, demapper takes channel observation Y and the inner *a priori* information A_1 and outputs soft values D_1. The extrinsic information $E_1 = D_1 - A_1$ is passed through bit-deinterleaver to become the *a priori* input A_2 for the outer decoder. The outer decoder feeds back the extrinsic information E_2 which becomes the *a priori* information A_1 for the inner decoder. Note that the variables $Y, D_1, A_1, E_1, D_2, A_2$ and E_2 denote log-likelihood ratios (L-values).

Having a noise corrupted channel observation Y and the *a priori* information A_1 as the input, the demapper then outputs extrinsic channel information E_1. The *a priori* information A_1 which comes from the feedback of the outer decoder E_2 are almost Gaussian distributed [59]. Moreover, large interleaver length keep the *a priori* information A_1 fairly uncorrelated over many iteration. Hence, it seems appropriate to model the *a priori* information A_1 as an independent Gaussian random variable n_{A_1} with variance $\sigma_{A_1}^2$ and zero mean. Together with the known transmitted information bits $x \in \{\pm 1\}$, *apriori* information A_1 can be written as :

$$A_1 = \mu_{A_1} x + n_{A_1}. \tag{3.29}$$

Since A_1 is based on Gaussian distribution, it is shown in [59] that $\mu_{A_1} = \sigma_{A_1}^2/2$, and thus its conditional pdf is :

$$p_{A_1}(\xi|X=x) = \frac{e^{-\frac{\left(\xi - \frac{\sigma_{A_1}^2}{2}x\right)^2}{2\sigma_{A_1}^2}}}{\sqrt{2\pi}\sigma_{A_1}}. \tag{3.30}$$

To measure the information content of the *a priori* knowledge, mutual information $I_{A_1} = I(X; A_1)$

3.2 Bit Interleaved Coded Modulation with Iterative Decoding (BICM-ID)

between the transmitted inner information bits X and the *a priori* information A_1 is used [59].

$$I_{A_1} = \frac{1}{2} \cdot \sum_{x=-1,1} \int_{-\infty}^{+\infty} p_{A_1}(\xi|X=x)$$
$$\times \log_2 \frac{2 \cdot p_{A_1}(\xi|X=x)}{p_{A_1}(\xi|X=-1) + p_{A_1}(\xi|X=+1)} d\xi, \quad 0 \leq I_{A_1} \leq 1. \quad (3.31)$$

Combining (3.30) and (3.31), we obtain [59] :

$$I_{A_1}(\sigma_{A_1}) = 1 - \int_{-\infty}^{+\infty} \frac{e^{-\frac{(\xi - \sigma_{A_1}^2/2)^2}{2\sigma_{A_1}^2}}}{\sqrt{2\pi}\sigma_{A_1}} \cdot (1 - \log_2[1 + e^{-\xi}]) d\xi. \quad (3.32)$$

The function $I_{A_1}(\sigma_{A_1} = \sigma)$, $\sigma > 0$ is monotonically increasing whose limit values are given as:

$$\lim_{\sigma \to 0} I_{A_1}(\sigma_{A_1} = \sigma) = 0, \quad \lim_{\sigma \to \infty} I_{A_1}(\sigma_{A_1} = \sigma) = 1, \quad \sigma > 0. \quad (3.33)$$

Mutual information is also used to quantify the extrinsic output $I_{E_1} = I(X; E_1)$. The demapper transfer characteristic is defined as a function of *a priori* bitwise mutual information I_{A_1} and E_b/N_0-value [59] :

$$I_{E_1} = T_1(I_{A_1}, E_b/N_0) \quad (3.34)$$

and it can be solved most conveniently by using Monte Carlo simulation.

The outer decoder transfer characteristic is defined as [59] :

$$I_{E_2} = T_2(I_{A_2}) \quad (3.35)$$

which describes the input/output relationship between the outer coded input A_2 and the outer coded extrinsic output E_2. Note that it is not depending on the E_b/N_0-value. I_{E_2} can be computed by assuming A_2 to be Gaussian distributed and applying the same equations as those for calculating T_1.

To visualize the exchange of extrinsic information, both decoder characteristics are plotted into a single diagram. This diagram is called EXIT chart. On the ordinate, the inner extrinsic and channel output I_{E_1} becomes the outer *a priori* input I_{A_2} (interleaving does not change the mutual information). On the abscissa, the outer extrinsic output I_{E_2} becomes the inner *a priori* input I_{A_1}.

3.2.5 Interleaving

The interleaver design is another critical parameter for high performance of BICM. The main objectives of the bit interleavers are to increase the minimum Euclidean distance between any two coded sequence as well as to mitigate the error propagation during the iterative decoding.

To reach this objectives, the key idea in designing a good interleaver for a BICM-ID system is to make the interleaved coded bits of a symbol as far apart as possible. This can be done by using many methods. In this thesis, we use a method called pseudorandom bit interleaver, which is presented in [25]. The design rules for pseudorandom bit interleaver are given as follow [25] :

- **Modularity** : The bit positions before and after interleaving must have the same modulo-m value. This ensures that the coded bits with different protection are distributed uniformly along the trellis due to their different positions at the channel-symbol labels.

- *Reverse Spread* : The m bits going to the same channel-symbol must be spread out as far apart as possible. This ensures feedback independence in bit metric recalculation and mitigation of the error propagation through iterative decoding.

Another important parameter regarding the bit interleaver is the block length of the interleaver. It is well known for many iterative decoding that the longer the interleaver, the better the error performance becomes. It is also the case for BICM-ID systems. A small block size of interleaver can cause a substantial degradation of BICM, because the probability that the bits coming from the same symbol still near to each other are still high. However, increasing the size of interleaver after a certain high SNR does not improve the performance of BICM-ID system any further.

Chapter 4

Multidimensional BICM-ID

Wei [27] and Pietrobon [28] showed that with MD modulation the error performance of Trellis Coded Modulation (TCM) systems can be significantly improved. This idea of MD Modulation was applied for MIMO transmission and BICM-ID by Simoens et al. in [29]. In [30] and [32] szczecinski et al. and schreckenbach et al. proposed modulation doping to improve the performance of BICM-ID in the low SNR regions which results in performance degradation at the error floor region. In this chapter, we propose a design for MD-BICM-ID which outperforms the two dimensional case in the whole SNR region. We use a combination of optimum MD labeling, modulation doping and designed intreleaver to outperform the two dimensional BICM-ID. We apply the technique on 8-PSK constellations.

This chapter is organized as follows. In section 4.1 a review of the MD-BICM-ID system is presented. Then we explain the RTS algorithm and the MD labeling in section 4.2. In section 4.3 we show how the modulation doping is implemented for the MD case. Interleaver design for MD-BICM-ID is presented in section 4.4. In section 4.5 simulation results for the proposed MD labeling compared to the two dimensional labeling are presented.

4.1 MD-BICM-ID Transmission System

Multidimensional signal mapping for BICM-ID considers several channel symbols over a multiple symbol interval [9]. This results in a large constellation space for the signal mapping construction compared to conventional 2D signal mapping. In 2D signal mapping each m bits are grouped and mapped to an M-ary constellation signal (where $M = 2^m$). When 4D signal mapping is used, each $2m$ bits are mapped to two M-ary constellation symbols. Figure 4.1 shows the 2D signal mapping and 4D signal mapping schemes with 8-PSK constellation. In 2D signal mapping, each 3 bits are grouped and mapped to one channel symbol. In 4D signal mapping, each 6 bits are mapped to two channel symbols.

As already known, for conventional BICM-ID with 8-PSK constellation, the optimized mapping is "Semi Set-Partitioning" (shown in Figure 3.4). Since 8! = 40320, this optimal mapping can be found easily through exhaustive search. For the multidiminsional case which we consider the search complexity is too high which means exhaustive search can not be used for this case. As discussed in chapter 3, the signal mapping problem can be analogized to QAP problem. This means all what we need is an algorithm to solve the QAP problem.

33

4 Multidimensional BICM-ID

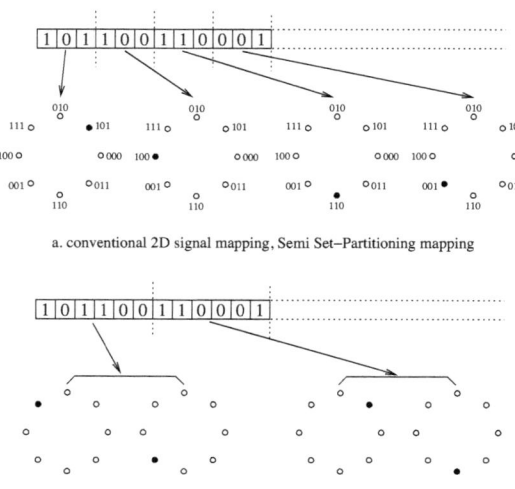

a. conventional 2D signal mapping, Semi Set–Partitioning mapping

b. multidimensional signal mapping, random mapping

Figure 4.1: 2D signal mapping versus 4D signal mapping

4.2 Reactive Tabu Search Algorithm

The Reactive Tabu Search (RTS) algorithm is a realization of Reactive Search (RS) principles in the framework of Tabu Search (TS) algorithm, where both of them originated from the classical Local Search (LS) algorithm.

The classical LS algorithm can be described as follows. From the generated initial solution, LS slightly modifies the configuration of this solution to get a better solution, i.e. the one with less cost function f value, by "looking in the neighborhood". LS can be effective if the neighborhood structure matches the characteristic of the problem, but it stops when the current configuration is a *local minimizer*, i.e. when all neighbors have higher f values.

Some possible actions to go beyond the local minima while aiming at better suboptimal points have been proposed, for example the repetition of LS restarting from different points, Simulated Annealing, Tabu Search and Genetic Algorithm [46]. However, there are some drawbacks of implementations of these schemes:

- **parameter tuning** : the user is required to carefully select the parameters for competitive results. This can be achieved either by a deep knowledge of the problem structure or simply a lengthy "trial and error" process, that is not always fully reproducible.

- **search confinement** : the points near the attraction basin after a *local minimizer* is encountered lose any interest of optimization. This cause an excessive computing time. To avoid it, *diversification* (exploration) should be activated. On the other hand, in the assumptions that neighbors have correlated cost function f, some effort should be spent in searching for better points located close to the just found local minimum point. This is called *intensification* (exploitation). These two requirements are conflicting. So, the proper balance between them is a crucial issue.

These drawbacks can be overcome by applying RS schemes, a generic scheme that introduce the past history of the search as the feedback to complement LS algorithm, where the past history of the search is used for :

- **feedback-based parameter tuning :** the algorithm maintain the internal flexibility needed to cover many problems, but the tuning is automated, and is executed while the algorithm runs and monitors its past behavior.
- **automated balance of diversification and intensification :** an automated heuristic balance can be obtained through feedback mechanisms, for example by starting with intensification, and by progressively increasing the amount of diversification only when there is evidence that diversification is needed.

As mentioned before, the Tabu Search (TS) algorithm is one of the proposed algorithm to make possible the action to go beyond the *local minimizer* of the LS algorithm. The method used by the TS algorithm to go beyond the *local minimizer* is through prohibition of selected moves available at the current point, which is stored in the data structure called *tabu_list*, and the enforcement of appropriate amounts of diversification to avoid that the search trajectory remains confined near a given local minimizer. This intelligent use of past history of the search to influence the future steps is the main competitive advantage of TS algorithm with respect to the other LS-based algorithms. However, there are some problems arising in TS algorithm, namely [46] :

- determination of an appropriate length of *tabu_list* for different task
- the robustness of the technique for a wide range of different problems
- the adoption of minimal computational complexity algorithms for using the search history

The use of RS principles in a framework of Tabu Search (TS) algorithm, which is called Reactive Tabu Search (RTS) algorithm, promises an attractive method to deal with these problems. These methods are described in the following part.

Self-adjusted prohibition method

The RTS algorithm has a simple mechanism to adapt the $list\_size$ of the TS algorithm to the properties of the optimization problem, namely through feedback (reactive) mechanism during the search. The $list\_size$ is equal to one at the beginning (the inverse of the given move is prohibited only at the next step), it increases only when there is evidence that diversification is needed, it decreases when this evidence disappears. In detail, the evidence that diversification is needed is signaled by the repetition of the previously visited configurations. All configuration found during the search are stored in memory. After a move is executed, the algorithm checks whether the current configuration has already been found and it reacts accordingly, i.e. $list\_size$ increases if a configuration is repeated, and $list\_size$ decreases if no repetition occurs during a sufficiently long period.

The escape mechanism

The TS algorithm is not sufficient to avoid long cycles (endless repetition of a sequence of configurations during the search). To increase the robustness of the algorithm, RTS algorithm uses a radical diversification step called *escape* mechanism. The *escape* mechanism is triggered when too many configurations are repeated too often. A simple *escape* mechanism consists of a number of random steps executed starting from the current configuration with a bias toward steps that bring the trajectory away from the current search region.

Fast algorithm for using the search history

In order to decrease the computational complexity of the searching process, the storage and access of the past events in RTS is executed through the well-known *hashing* or *digital-tree* techniques which results in an approximately constant CPU time with respect to the number of iteration. Therefore, the overhead caused by the use of history is negligible for the tasks requiring a non trivial number of operations to evaluate the cost function in the neighborhood.

Having the idea about the basic functions and memory structure of RTS algorithm, in algorithm 1 we presents the basic RTS loop in the *pseudo code* form.

Algorithm 1 Basic RTS Loop [46]

```
  procedure initialization
  begin
  Initialize max_time
  Initialize report_every
  end
  while Task.current.time ¡ max_time do
    begin
    if check_for_repetitions(&Task.current) == DO_NOT_ESCAPE) then
        evaluate_neighborhood();
        choose_best_move();
        make_tabu(Task.chosen_bit);
        update_current_and_best();
    else
        escape();
    end if
    if ((Task.current.time % report_every) == 0 ) then
        PrintStateCompressed()
    end if
    end
  end while
```

Optimized signal mapping for MD-BICM-ID system

For 4-dimensional BICM-ID system, equation (3.25) is modified to:

$$\min_{\mu} \sum_{k=1}^{2m} \sum_{b=0}^{1} \sum_{\mathbf{X} \in \chi_b^k} \frac{1}{\| x_1 - \tilde{z}_1 \|^2 + \| x_2 - \tilde{z}_2 \|^2} \tag{4.1}$$

the "distance" between multidimensional signal points $\mathbf{X}_i$ and $\mathbf{X}_j$ becomes:

$$d_{\mathbf{X}_i, \mathbf{X}_j} = \begin{cases} \frac{1}{\|x_{i1} - x_{j1}\|^2 + \|x_{i2} - x_{j2}\|^2}, & i \neq j \\ 0, & i = j \end{cases} \tag{4.2}$$

4.2 Reactive Tabu Search Algorithm

and the "flow" between labels i and j remains the same as that given in equation (3.27). Finally, the QAP problem of multidimensional signal mapping can be written as:

$$\min_{\mu} \sum_{i=0}^{2^{2m}-1} \sum_{j=0}^{2^{2m}-1} f_{ij} d_{\mu(i),\mu(j)} \qquad (4.3)$$

Reference [18] provides the RTS algorithm as an open source code for application of the QAP problem. We use this software to find the optimized signal mapping for MD-BICM-ID with 8-PSK constellation.

Optimized signal mapping for MD-BICM-ID system with 8-PSK modulation

Figure 4.2 illustrates the 4D signal with 8-PSK constellation.

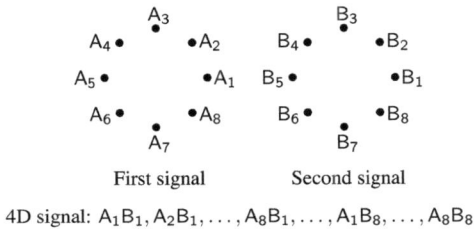

4D signal: $A_1B_1, A_2B_1, \ldots, A_8B_1, \ldots, A_1B_8, \ldots, A_8B_8$

Figure 4.2: 4-dimensional 8-PSK signal constellation

The RTS algorithm is initialized with a random constellation labeling (configuration). Then it will search configurations in its neighborhood that have lower fitness values. These configurations are made tabu for a certain period of time.

By applying the RTS algorithm [17], we obtained the optimized 4D 8PSK constellation labeling. This labeling is presented in Table 4.1. For convenience, we present the optimized constellation label in decimal form, each of it represents 6 bits. The label sequence is based on the sequence presented in Figure 4.2, namely the horizontal axis represents the first transmit symbol and the vertical axis represents the second one.

Table 4.1: The optimized signal mapping for 4D BICM-ID system with 8-PSK constellation.

	A_1	A_2	A_3	A_4	A_5	A_6	A_7	A_8
B_1	15	30	23	18	27	10	3	6
B_2	45	60	53	48	57	40	33	36
B_3	5	20	29	24	17	0	9	12
B_4	59	42	35	38	47	62	55	50
B_5	19	2	11	14	7	22	31	26
B_6	49	32	41	44	37	52	61	56
B_7	25	8	1	4	13	28	21	16
B_8	39	54	63	58	51	34	43	46

4 Multidimensional BICM-ID

4.3 MD-BICM-ID with Modulation Doping

Modulation doping uses two different mappings (e.g. Gray and anti-Gray) in one transmission block [30]. Such design provide optimum matching between the labeling and the CC of BICM-ID system without exhaustive search for the suitable signal mapping. Such scheme does not increase the receiver complexity. Here, we use modulation doping in the 4D BICM-ID systems. Given that the error floor of the 4D BICM-ID is much lower than that with 2D labeling the modulation doping will have less impact on the performance at the high SNR region in the 4D case.

Given a coded interleaved block with length N, doping ratio α, signal mapping scheme μ_1 and μ_2. The first $N \times \alpha$ bits in the block will be mapped to the channel symbols using mapping μ_1, and the rest bits using mapping μ_2. Figure 4.3 shows the case in 4D BICM-ID with 8-PSK modulation. The first mapping μ_1 is an optimal mapping with lowest error floor obtained in Section 4.2 using the RTS algorithm. The other mapping μ_2 is conventional Gray mapping. Note that the bits with different mappings are spread throughout the whole block due to the bit interleaver.

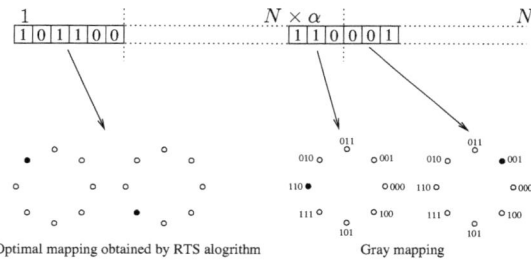

Figure 4.3: Doping technique for MD-BICM-ID with 8-PSK modulation

The modulation doping uses two already known mappings and the doping ratio to adapt the system to a desirable performance. Such scheme is easy to tune and does not increase the complexity of the receiver. By increasing the doping ratio the system can reach a lower error floor at high SNR. However, the iterative decoding does not work at low SNR. The selection of the doping ratio depends on target BER and channel quality. EXIT chart is a useful tool to select the desired doping ratio for a given outer code.

4.4 Interleaver Design for MD-BICM-ID

The interleaver design is another important part of system design for a iteratively decoded systems. For short blocks, the design of the interleaver is a critical problem. In BICM-ID systems, the main function of the bit interleaver is to break the correlation of the coded bits and increase the minimum Euclidean distance between any two code sequences to mitigate the error propagation during the iterative decoding.

Consider a 4D BICM-ID system with M-ray modulation, interleaver size N. we propose the following interleaver construction:

- *Step 1*: In order to separate these $2m$ coded bits as apart as possible, a $4m \times (N/4m)$ matrix is constructed. The coded bits are written column-wise into the matrix.

4.4 Interleaver Design for MD-BICM-ID

- *Step 2* : For each row, an S-random interleaver is used to introduce the "randomness" and ensure that the neighbour channel symbols are mapped from coded bits which are far enough. Such design can mitigate the affect of a burst decoding errors. Each row should use a different S-random interleaver. Usually choosing $S < \sqrt{N/2}$ produces a solution in acceptable time [24].

- *Step 3* : The bits are read row-wise from the matrix. In order to ensure that the $2m$ bits of each group in the original coded sequence are separated by at least $N/4m$ bits after interleaving, we separate these $4m$ rows into two parts. 1 to $2m$ rows are part A, $2m+1$ to $4m$ row are part B. We read one row from part A first, then one row from part B, one row from part A, and so on.

Figure 4.4 illustrates this interleaver design.

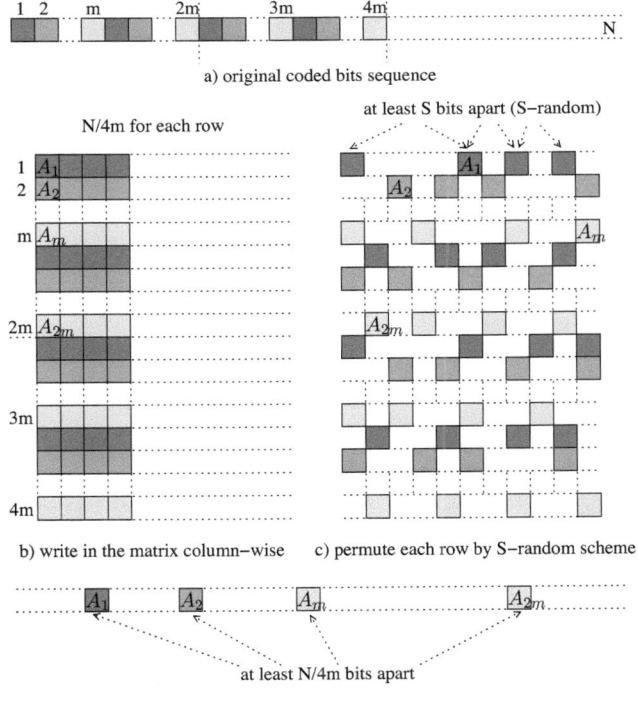

Figure 4.4: Interleaver design for 4D BICM-ID with M-ary ($M = 2^m$) modulation, interleaver size is N bits

4.5 Simulation of the MD-BICM-ID

We use an 8-state, rate 2/3 convolutional code as the channel encoder, which has $(41, 30, 75)_8$ representation in [19] (generator sequences $\mathbf{g}_1 = (4, 2, 6)_8$ and $\mathbf{g}_2 = (1, 4, 7)_8$ [34]). And each coded block contains 4000 information bits. This is the same setup used in [25].

The error performances of 2D and 4D BICM-ID with 8-PSK modulation over Rayleigh fading channel are presented in Figure 4.5.

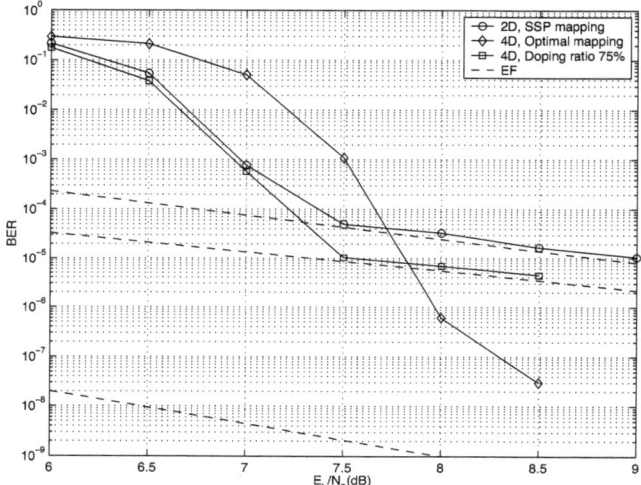

Figure 4.5: Performance comparison between 2D, 4D BICM-ID and 4D BICM-ID with doping over Rayleigh fading channel. 8-state, R=2/3 convolutional code, 8-PSK modulation, 4000 information bits/block and 8 iterations are used.

It is known that in conventional 2D BICM-ID with 8-PSK modulation, SSP mapping [25] has the best performance due to the biggest harmonic mean Euclidean distance $\tilde{d}_h^2$ with ideal feedback. When 4D BICM-ID with optimized mapping, which we got by the RTS algorithm and presented in Table 4.1, is used, the system outperforms the 2D BICM-ID system at high SNR and a much lower error floor can be achieved. More specifically, 4D BICM-ID gets 1.2 dB gain compared to 2D BICM-ID at $BER = 10^{-5}$. However, there is a performance degradation at low SNR.

The convergence behaviour of the BICM-ID system can be well explained by the use of EXIT chart. Figure 4.6 shows the EXIT chart of 4D BICM-ID with 8-PSK modulation over a Rayleigh fading channel. At $E_b/N_0 = 6.5$ dB, the tunnel between demapper and outer convolutional decoder is closed. This means that the iterative decoding does not work at this SNR. And at $E_b/N_0 = 8$ dB, the tunnel opens widely, which means the BER curve can converge to the asymptotic error performance after a few iterations at this SNR.

The degradation in performance of 4D BICM-ID at low SNR can be eliminated by using *modulation doping* as presented in Section 4.3. Figure 4.7 shows the BER performance of 4D BICM-ID with different doping ratios. Lower doping ratio yields better performance at low SNR region, while the per-

4.5 Simulation of the MD-BICM-ID

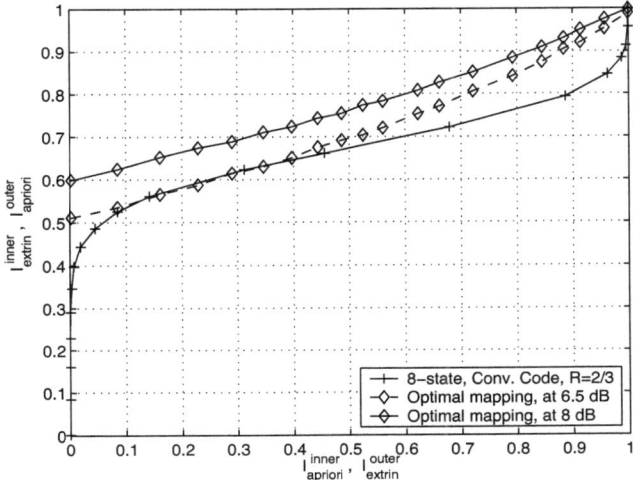

Figure 4.6: EXIT chart of 4D BICM-ID with 8-PSK modulation over Rayleigh fading channel. E_b/N_0 = 6.5 and 8 dB

formance at high SNR region is degraded. It is seen that if we target the $BER = 10^{-5}$, doping ratio = 75% is appropriate (also plotted in Figure 4.5 for comparison). It provides a better performance than the 2D mapping in the whole SNR region. More specifically, the 4D BICM-ID with doping ratio 75% outperforms the 4D BICM-ID without doping by 0.3 dB at $BER = 10^{-5}$.

MD-BICM-ID with doping technique

Figure 4.7 shows the effect of different doping ratios of 4D BICM-ID. In order to improve the error performance at low SNR, modulation doping technique is used. Lower doping ratio yields better performance at low SNR region, while the performance at high SNR region is degraded. By careful tuning with the doping ratio, a desired system can be got. If we target the $BER = 10^{-5}$, doping ratio = 75% is appropriate. At this BER, the 4D BICM-ID with doping ratio 75% outperforms the 4D BICM-ID without doping by 0.3 dB.

The EXIT chart of 4D BICM-ID with doping technique at $E_b/N_0 = 7.5$ dB is shown in Figure 4.8. Different doping rates lead to different slopes of the EXIT curves. Before intensive simulation, we can use the EXIT chart to estimate the error performance and select suitable parameters for the simulation. In this case, doping rate = 75% is suitable for us.

The effect of Interleaver design for BICM-ID

The error floor becomes very high after using modulation doping compared to 4D BICM-ID. Here a designed interleaver could lower the error floor. We use the design method introduced in Section 4.4 and choose $S = 13$ as the parameter of the *S-random* interleaver. Figure 4.9 shows that our interleaver

4 Multidimensional BICM-ID

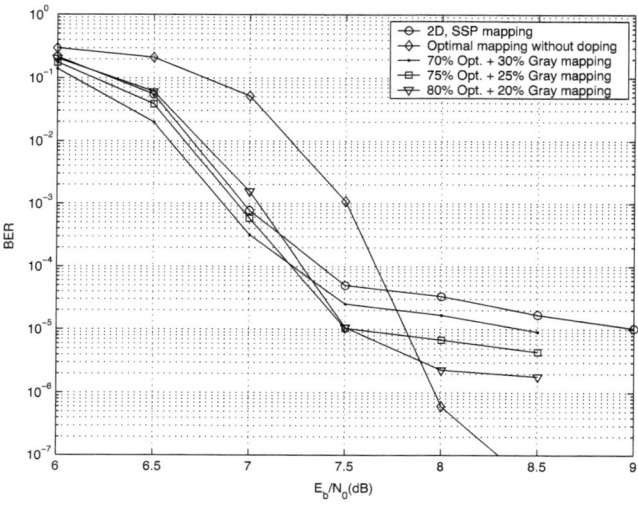

Figure 4.7: Performance comparison between different doping ratios of 4D BICM-ID over Rayleigh fading channel.

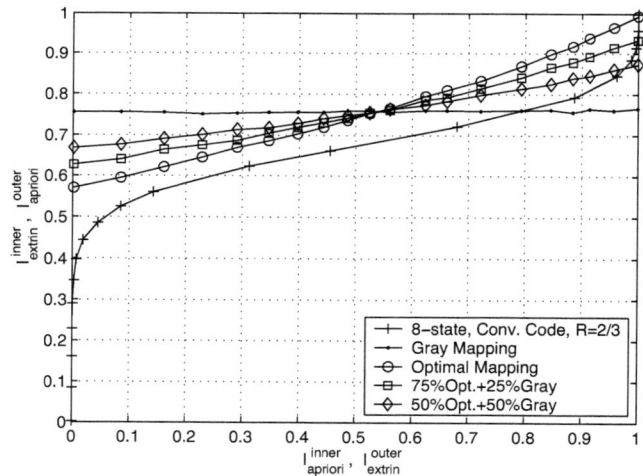

Figure 4.8: EXIT chart of MD-BICM-ID with doping technique at $E_b/N_0 = 7.5dB$

design improves the performance by 0.2 dB at $BER = 10^{-5}$ compared to using random interleavers and outperforms the 4D BICM-ID without doping by 0.5 dB at $BER = 10^{-5}$.

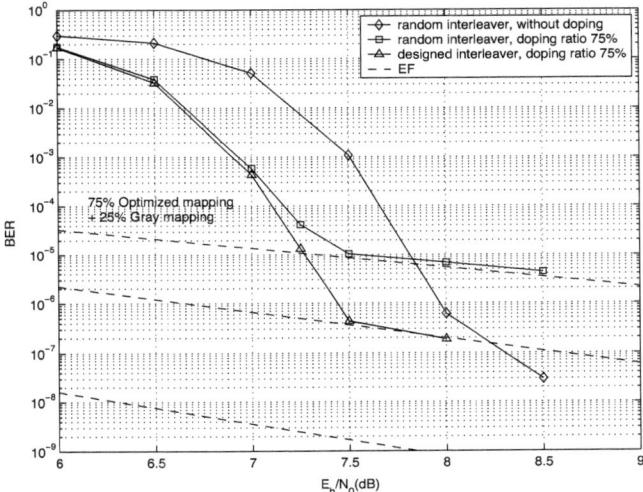

Figure 4.9: BER performance comparison with random interleaver and designed interleaver.

Chapter 5

Space Time Signaling

The information theoretic optimality of the Alamouti scheme for the special case of one receive antenna [63], [14] motivates the use of this scheme as an inner code for concatenated space time coding [10], [64].

In MIMO-OFDM transmission there is frequency diversity in addition to space diversity, resulting from the frequency selectivity of the channel. The design of Space Time Frequency (STF) codes can be greatly simplified by designing groups of STF codes that achieve full frequency and space diversity. For a transmission over Rayleigh fading channels the key point to increase diversity gain is to apply a certain rotation to a classical signal constellation in such a way that any two points achieve the maximum number of distinct components [66].

In this chapter the simple case of two transmit, one receive antennas is considered. In the first part we consider two subcarriers for each transmit antenna. Assuming uncorrelated space and frequency channel fading coefficients and we construct a $2 \times 2 \times 2$ STF code that achieves a diversity order of 4. It is designed by concatenating an Alamouti scheme with a real constellation rotation. Each one of the rotated constellation points is sent through a different frequency band in order to guarantee frequency diversity. In the second part we consider the extension of the MD labeling for the Space Time case considering the Alamouti scheme as the inner space time code and we construct a Multidimensional Bit Interleaved Space Time Coded Modulation with Iterative Decoding (MD-BISTCM-ID).

5.1 A Simple Full-Rate Full-Diversity Space Time Frequency Transmission Scheme

We consider the transmission scheme depicted in Figure 5.1 [6]. The constellation points are divided into two column vectors $\mathbf{c}_1 = \begin{bmatrix} c_{11} & c_{12} \end{bmatrix}^T$ and $\mathbf{c}_2 = \begin{bmatrix} c_{21} & c_{22} \end{bmatrix}^T$ with $c_{ij} \in \mathcal{A}$, where $\mathcal{A}$ is a set of constellation points (e.g. M-QAM). Each column vector is rotated using the real rotation matrix:

$$\mathbf{M} = \begin{bmatrix} cos(\theta) & sin(\theta) \\ -sin(\theta) & cos(\theta) \end{bmatrix} \quad (5.1)$$

where θ is the rotation angle to be optimized in section III. The resulting rotated constellation is given by

$$\mathbf{x}_i = \mathbf{M} \cdot \mathbf{c}_i, i \in \{1, 2\}, \quad (5.2)$$

where $\mathbf{x}_1 = \begin{bmatrix} x_{11} & x_{12} \end{bmatrix}^T$ and $\mathbf{x}_2 = \begin{bmatrix} x_{21} & x_{22} \end{bmatrix}^T$ with $x_{ij} \in \mathcal{C}$. An orthogonal space time code

5.1 A Simple Full-Rate Full-Diversity Space Time Frequency Transmission Scheme

is constructed from $\mathbf{x}_1$ and $\mathbf{x}_2$ [3] with the code

$$\begin{bmatrix} x_{11}^* & -x_{21}^* \\ x_{21} & x_{11} \end{bmatrix} \text{ transmitted over } f_1 \qquad (5.3)$$

and

$$\begin{bmatrix} -x_{12}^* & x_{22}^* \\ x_{22} & x_{12} \end{bmatrix} \text{ transmitted over } f_2. \qquad (5.4)$$

The rows in (5.3) and (5.4) represent time slots and the columns represent antennas.

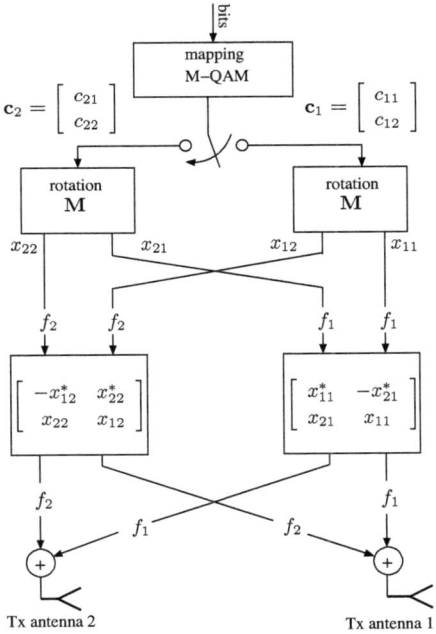

Figure 5.1: The STF transmitter.

The two symbols in each Alamouti code are linked through rotation with the two symbols of the other Alamouti code to guarantee frequency diversity. The symbols are then OFDM modulated and sent through the two antennas.

5.1.1 The STF detector

The detector is depicted in Figure 5.2. In this Figure the indices i and j of the fading coefficients h_{ij} represent the transmit antenna and the frequency band, respectively. The indices i and j of the of the received signals r_{ij} and the noise n_{ij} represent the time slot and the frequency band, respectively. The

45

5 Space Time Signaling

received signals $\mathbf{r}_1 = \begin{bmatrix} r_{11} & r_{21} \end{bmatrix}^T$ and $\mathbf{r}_2 = \begin{bmatrix} r_{12} & r_{22} \end{bmatrix}^T$ at the input of the two combiners are given by:

$$\begin{aligned} r_{11} &= h_{11}x_{11} + h_{21}x_{21} + n_{11} \\ r_{21} &= -h_{11}x_{21}^* + h_{21}x_{11}^* + n_{21} \end{aligned} \tag{5.5}$$

for combiner one and ,

$$\begin{aligned} r_{12} &= h_{22}x_{22} + h_{12}x_{12} + n_{12} \\ r_{22} &= -h_{22}x_{12}^* + h_{12}x_{22}^* + n_{22} \end{aligned} \tag{5.6}$$

for combiner two.
The combining scheme builds the following combined signals [3]:

$$\begin{aligned} \tilde{x}_{11} &= h_{11}^* r_{11} + h_{21} r_{21}^* \\ \tilde{x}_{21} &= h_{21}^* r_{11} - h_{11} r_{21}^* \end{aligned} \tag{5.7}$$

and ,

$$\begin{aligned} \tilde{x}_{12} &= h_{12}^* r_{12} - h_{22} r_{22}^* \\ \tilde{x}_{22} &= h_{22}^* r_{12} + h_{12} r_{22}^* \end{aligned} \tag{5.8}$$

Substituting this in (5.5) and (5.6) we get

$$\begin{bmatrix} \tilde{x}_{11} \\ \tilde{x}_{21} \end{bmatrix} = \begin{bmatrix} |h_{11}|^2 + |h_{21}|^2 & 0 \\ 0 & |h_{11}|^2 + |h_{21}|^2 \end{bmatrix} \cdot \begin{bmatrix} x_{11} \\ x_{21} \end{bmatrix} + \begin{bmatrix} h_{11}^* n_{11} + h_{21} n_{21}^* \\ h_{21}^* n_{11} - h_{11} n_{21}^* \end{bmatrix} \tag{5.9}$$

and

$$\begin{bmatrix} \tilde{x}_{12} \\ \tilde{x}_{22} \end{bmatrix} = \begin{bmatrix} |h_{12}|^2 + |h_{22}|^2 & 0 \\ 0 & |h_{12}|^2 + |h_{22}|^2 \end{bmatrix} \cdot \begin{bmatrix} x_{12} \\ x_{22} \end{bmatrix} + \begin{bmatrix} h_{12}^* n_{12} - h_{22} n_{22}^* \\ h_{22}^* n_{12} + h_{12} n_{22}^* \end{bmatrix} \tag{5.10}$$

The four signals are separated completely. However, to benefit from frequency diversity with minimum complexity we detect each two of the rotated pairs jointly. By rearranging (5.9) and (5.10) we get:

$$\begin{bmatrix} \tilde{x}_{11} \\ \tilde{x}_{12} \end{bmatrix} = \begin{bmatrix} |h_{11}|^2 + |h_{21}|^2 & 0 \\ 0 & |h_{12}|^2 + |h_{22}|^2 \end{bmatrix} \cdot \begin{bmatrix} x_{11} \\ x_{12} \end{bmatrix} + \begin{bmatrix} h_{11}^* n_{11} + h_{21} n_{21}^* \\ h_{12}^* n_{12} - h_{22} n_{22}^* \end{bmatrix} \tag{5.11}$$

5.1 A Simple Full-Rate Full-Diversity Space Time Frequency Transmission Scheme

$$\begin{bmatrix} \tilde{x}_{21} \\ \tilde{x}_{22} \end{bmatrix} = \begin{bmatrix} |h_{11}|^2 + |h_{21}|^2 & 0 \\ 0 & |h_{12}|^2 + |h_{22}|^2 \end{bmatrix} \cdot \begin{bmatrix} x_{21} \\ x_{22} \end{bmatrix} + \begin{bmatrix} h_{21}^* n_{11} - h_{11} n_{21}^* \\ h_{22}^* n_{12} + h_{12} n_{22}^* \end{bmatrix} \quad (5.12)$$

Using (2) in (5.11), then the ML decoding rule for (5.11) is to choose the constellation points $\mathbf{c}_1 = \begin{bmatrix} c_{11} & c_{12} \end{bmatrix}^T$ that minimize the sum of the squared Euclidean distance:

$$d^2(\tilde{x}_{11}, (|h_{11}|^2 + |h_{21}|^2)(cos(\theta)c_{11} + sin(\theta)c_{12})) + \\ d^2(\tilde{x}_{12}, (|h_{12}|^2 + |h_{22}|^2)(-sin(\theta)c_{11} + cos(\theta)c_{12})) \quad (5.13)$$

Similarly for (5.12). From (5.13) it is clear that for high SNR only deep fading in all the four fading coefficients will cause signal loss which guarantees a diversity of 4. For higher order constellations a sphere detector can be used instead of the ML-detector which results in a reduced complexity [67], [68].

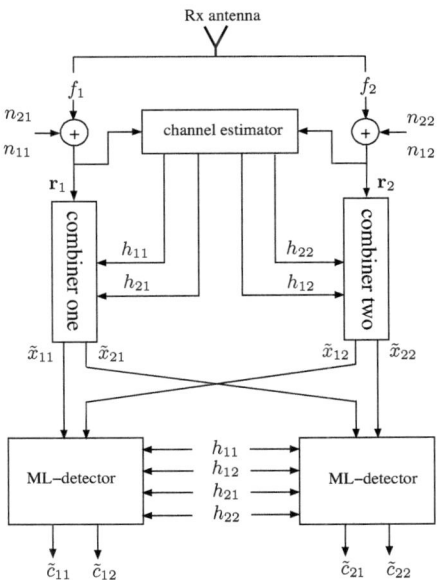

Figure 5.2: The STF detector.

5.1.2 Optimum Rotations

The performance of the transmitter in Figure 5.1. is a function of one parameter which is the rotation angle θ. As in [66] for the case of rotated constellation transmission over Rayleigh channel the rotation angle for the optimum performance of the whole system can be found. However, instead of searching for the optimum θ which maximizes the minimum product distance we can bypass the design criteria by searching for the optimum θ that minimizes the average error probability at the output of the ML-detector. This can be done by simulating the BER curve of the system at a moderate SNR value. The BER of the STF code is plotted in Figure 5.3 with QPSK constellation points at SNR = 10 dB for $\theta = 0...\pi/4$. The optimum value is found to be 0.49. We noticed that this optimum value of θ is the same for other SNR values.

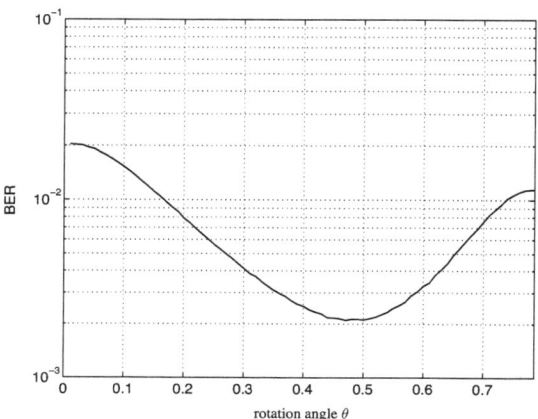

Figure 5.3: Performance for different values of θ at SNR=10dB.

5.1.3 Simulation Results

In simulations, we used a normalized QPSK constellation with average energy per symbol $E_s = 1$. The additive white Gaussian noise has a variance $\sigma^2 = m \cdot n/(2SNR)$ per real dimension, where $m = 2$ is the number of transmit antennas and $n = 2$ is the number of sub-carriers. Bit error probability curves are plotted as a function of SNR in dB. This normalization is done to enable comparison with the single antenna system and the Alamouti scheme without frequency diversity. The simulation results are shown in Figure 5.4. At 10^{-5} about 7.2 dB diversity gain is achieved in comparison with the Alamouti scheme. Additionally, we plotted the results for the STF code in [64] which uses complex precoders. A gain of about 0.8 dB is achieved by the proposed real rotations. In addition to the achieved gain, the real and imaginary parts of 5.11 and 5.12 can be decoded separately in the case of real rotation, which results in reduced decoding complexity.

5.2 Multidimensional BISTCM-ID Transmission

Due to its simplicity and information theoretic optimality [14], the Alamouti scheme [3] is used as an inner STC for two transmit and one receive antenna systems [6] [10]. BISTCM-ID with Alamouti scheme consists of concatenating a Convolutional Code (CC) with the Alamouti scheme separated by a

5.2 Multidimensional BISTCM-ID Transmission

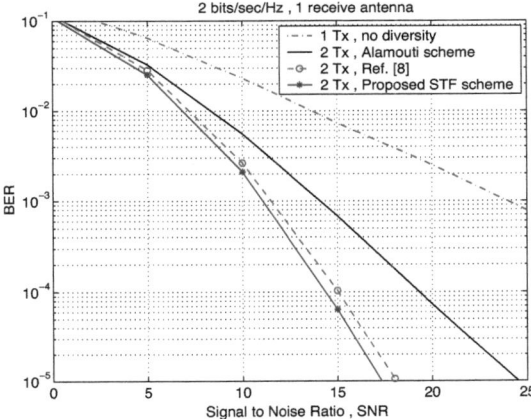

Figure 5.4: The proposed STF code compared to the Alamouti scheme and a STF code based on complex precoders.

bit-interleaver. The use of iterative decoding at the receiver side results in turbo gains. As mentioned in the previous chapters constellation labeling plays an important role in the asymptotic performance of such a system. In this section, we propose a method to improve the performance of BI-STCM-ID system by applying multidimensional labeling [7]. We apply the RTS algorithm to search for optimum multidimensional 16-QAM constellation for BISTCM-ID system which results in large asymptotic coding gains over two dimensional labeling proposed recently in [4].

The block diagram of BISTCM-ID system which uses the Alamouti scheme with one receive antenna is depicted in Figure 5.2, the CC with code rate k_c/n_c is first used to encode the information bit sequence $\underline{u}$. Then the output of the CC is bit-interleaved. The interleaved sequence $\underline{v}$ are mapped to signal points from an M-QAM constellation according to the labeling rule μ. Next, each two complex signals (x_t^1, x_t^2) are grouped to form a space-time codeword matrix $\mathbf{X}_t$ based on the Alamouti scheme. The channels considered are fast-fading channels, over which the Rayleigh coefficient remains constant during the transmission of one single codeword matrix $\mathbf{X}_t$. The received signal is given by

$$\mathbf{Y}_t = \mathbf{X}_t \mathbf{H}_t + \mathbf{N}_t. \tag{5.14}$$

where $\mathbf{N}_t$ is the noise matrix modeled as independent and identically distributed (iid) complex Gaussian random variables with zero mean and variance $N_0/2$ per dimension.

At the receiver, the CSI is assumed to be perfectly known. The extrinsic log-likelihood ratio (LLR) for each bit $\mathcal{L}_e(v_t^k; O)$ is first calculated using the maximum *a posteriori* probability (MAP) rule at the space-time demapper. Then, the bit-interleaved version of which is fed into the SISO decoder to produce the extrinsic LLR of each coded bit, which is used as *a priori* information for the next iteration [4].

The asymptotic error performance of the considered BISTCM-ID system over Rayleigh fading channel is derived from the BICM union bound with the assumption of "error-free feedback" [58] and can be

5 Space Time Signaling

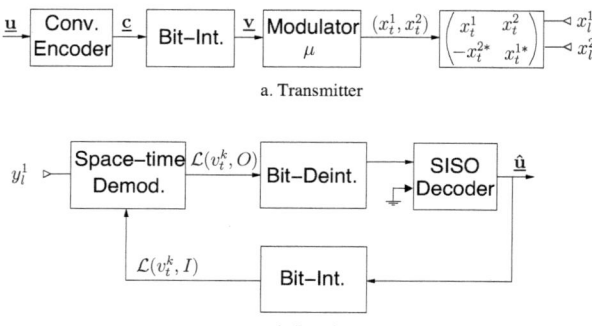

Figure 5.5: The Block Diagram of BISTCM-ID System.

written as

$$\log_{10} P_b \simeq -\frac{d_f}{5}\left[(R\Omega^2)_{dB} + \left(\frac{E_b}{N_0}\right)_{dB}\right] + \text{const}, \quad (5.15)$$

where R is the overall information rate and Ω^2 is the coding gain.

Equation (5.15) shows that at high SNR values, the asymptotic BER curve of BISTCM-ID in logarithmic scale approximates a straight line, where the slope is controlled by the diversity order of the system and the horizontal offset is provided by coding gain.

The coding gain of BISTCM-ID which employs the Alamouti scheme with 1 receive antenna is given by:

$$\Omega^2(\mu) = \left[\frac{1}{m2^m}\sum_{k=1}^{m}\sum_{b=0}^{1}\sum_{x\in\chi_b^k}(|x-\tilde{z}|^2)^{-2}\right]^{-1/2} \quad (5.16)$$

where x and $\tilde{z}$ are signals whose labels are different only in the k-th bit position and χ_b^k is the subset of signal constellation points whose label in the k-th bit position has the value of b. Obviously, constellation labeling plays an important role in the optimization of coding gain.

Since there are $(2^m)!$ possible combinations of constellation labeling for M-ary constellations, the design of constellation labeling results in a huge number of possible solutions. However, this problem is analogous to the facility-location problem in classical QAP [4], which can be solved by applying the generic solution of QAP.

By assuming the constellation points as locations and the constellation labeling as facilities, the optimization problem is to find the one-to-one mapping between the facilities and the locations which minimizes the assignment cost function

$$\min_{\mu}\sum_{k=1}^{m}\sum_{b=0}^{1}\sum_{x\in\chi_b^k}(|x-\tilde{z}|^2)^{-2}. \quad (5.17)$$

Equation (5.17) implies that we have to optimize the squared Euclidean distance between two signal points whose labels are different only in one bit position, among all members of all the subsets χ_b^k. Therefore, for analogy with facility-location problem of QAP, the "distance" between signal point x_i and

5.2 Multidimensional BISTCM-ID Transmission

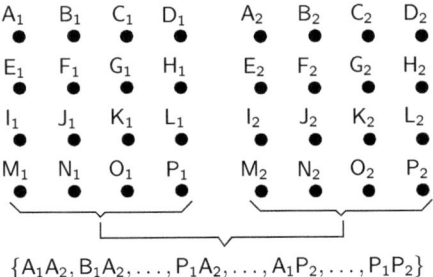

Figure 5.6: Transformation of two dimensional 16-QAM constellation space into multidimensional 16-QAM constellation space.

x_j is defined as [4]

$$d_{x_i,x_j} = \begin{cases} (|x_i - x_j|^2)^{-2}, & i \neq j \\ 0, & i = j \end{cases} \tag{5.18}$$

and the "flow" between constellation-point's label i and j is defined as

$$f_{ij} = \begin{cases} 1, & i \text{ and } j \text{ differ only in one bit position} \\ 0, & \text{otherwise} \end{cases} \tag{5.19}$$

Thus, the QAP formulation of mapping rule optimization can be written as

$$\min_{\mu} \sum_{i=0}^{2^m-1} \sum_{j=0}^{2^m-1} f_{ij} d_{\mu(i),\mu(j)}. \tag{5.20}$$

There are several algorithms that can be used to solve QAP, among which the RTS is considered to be one of the best and most efficient algorithms.

5.2.1 Multidimensional Constellation Labeling for the BISTCM-ID

For the considered BISTCM-ID system, each space-time codeword consists of two transmit symbols. Since the Alamouti scheme allows the transmission of two symbols in each space time codeword and the calculation of metrics at the receiver is based on the space-time matrix $\mathbf{X}_t$, multidimensional labeling that considers two consecutive transmit symbols will give more flexibility in the mapping rule design without increasing the decoding complexity.

Figure 5.6 illustrates the multidimensional constellation space. Obviously, there are $(2^4)^2 = 256$ vertices in the multidimensional constellation space.

As we did for the MD-BICM-ID case, we slightly modify the cost function of the QAP formulation of two-dimensional constellation labeling given in (5.17) by considering the two symbols in the formula, which results in

$$\min_{\mu} \sum_{k=1}^{2m} \sum_{b=0}^{1} \sum_{\mathbf{X} \in \chi_b^k} (|x_1 - \tilde{z}_1|^2 + |x_2 - \tilde{z}_2|^2)^{-2} \tag{5.21}$$

51

5 Space Time Signaling

Table 5.1: The proposed multidimensional 16-QAM constellation labeling for BISTCM-ID system with Alamouti scheme.

	A_1	B_1	C_1	D_1	E_1	F_1	G_1	H_1	I_1	J_1	K_1	L_1	M_1	N_1	O_1	P_1
A_2	10	130	117	253	6	142	251	247	72	192	49	61	66	78	185	181
B_2	0	12	127	213	46	132	87	223	96	108	59	21	106	68	55	191
C_2	139	163	118	248	135	141	218	242	201	225	50	56	195	77	152	176
D_2	3	9	254	244	15	129	214	229	99	105	186	52	75	65	182	188
E_2	34	136	121	241	170	166	115	217	226	204	25	145	202	198	179	157
F_2	40	160	81	93	36	172	91	211	232	228	19	31	102	238	155	151
G_2	39	169	122	112	175	165	82	88	231	237	26	16	207	197	146	28
H_2	43	33	84	124	5	45	94	208	235	101	22	62	71	111	158	148
I_2	30	150	69	239	20	190	107	103	92	212	1	47	80	126	41	37
J_2	144	156	79	199	60	180	67	203	240	252	11	7	120	116	35	171
K_2	159	183	70	234	149	189	74	98	221	245	2	42	209	125	8	32
L_2	147	153	206	230	29	177	194	200	243	249	138	38	89	113	162	168
M_2	18	154	109	229	58	54	97	233	210	222	13	133	90	86	161	173
N_2	24	178	193	205	48	184	73	227	216	246	131	143	114	250	137	167
O_2	23	187	110	100	63	53	64	104	215	255	14	4	95	85	128	44
P_2	27	51	196	236	17	57	76	224	219	119	134	174	83	123	140	164

where $(x_1, x_2) \in \mathbf{X}$ and χ_b^k in this case is the subset of constellation points in the multidimensional constellation space whose bit in k-th bit position has the value of b. Consequently, the "distance" between signal vertex $\mathbf{X}_i$ and $\mathbf{X}_j$ is modified to

$$d_{\mathbf{X}_i,\mathbf{X}_j} = \begin{cases} \left(|x_{i1} - x_{j1}|^2 + |x_{i2} - x_{j2}|^2\right)^{-2}, & i \neq j \\ 0, & i = j \end{cases} \quad (5.22)$$

whereas the "flow" between constellation-point's labeling i and j remains the same.

By applying the RTS algorithm [17], we obtained the optimized multidimensional 16-QAM constellation labeling for the considered multidimensional BISTCM-ID system. This labeling is presented in Table 5.1. For convenience, we present the optimized constellation label in decimal form, each of it represents 8 bits. The label sequence is based on the sequence presented in Figure 5.6, namely the horizontal axis represents the first transmit symbol and the vertical axis represents the second one.

Figure 5.7 shows the learning curve of the RTS algorithm when searching for the multidimensional labeling. It can be seen that the balance between intensification, diversification and escape mechanisms enables the RTS algorithm to go further to the minimum fitness function.

5.2.2 Simulation Results

The simulation environment used is the same as in [4]. The performance comparison of the considered multidimensional BISTCM-ID with Ritcey's labeling and the proposed multidimensional labeling is shown in Figure 5.2.2 8 iterations are performed. The proposed multidimensional labeling significantly outperforms Ritcey's labeling by about 4 dB asymptotic coding gain.

To clarify the advantage of the proposed labeling the EXIT chart for Ritcey's labeling and the new one at $E_b/N_0 = 6$ dB is plotted in Figure 5.9. The EXIT chart shows that the tunnel of the multidimensional labeling is wider open than the two dimensional one when the a priori information of the demapper is large. This explains why the multidimensional labeling performs better at high SNR when a large number of iterations are used.

5.2 Multidimensional BISTCM-ID Transmission

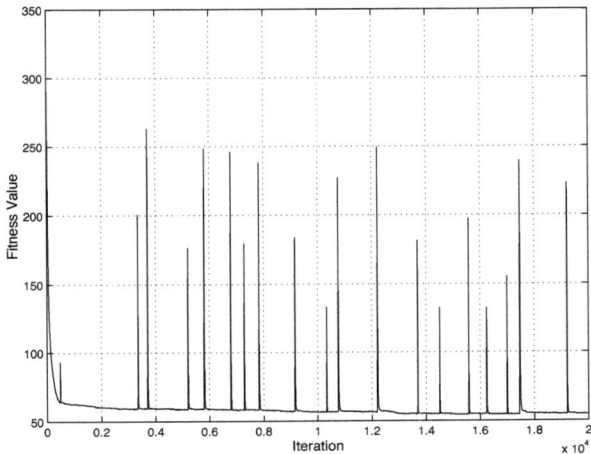

Figure 5.7: Learning curve of RTS algorithm.

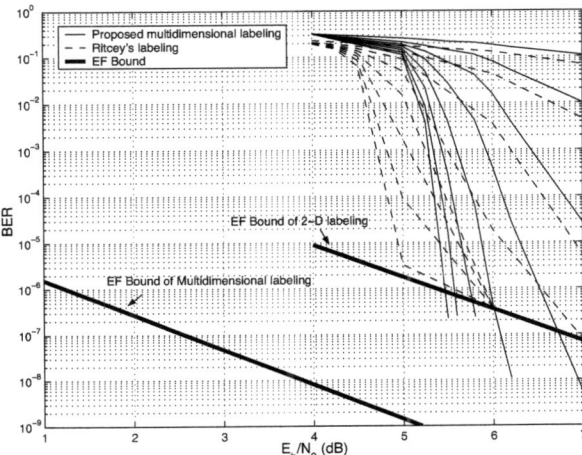

Figure 5.8: BER performance of multidimensional labeling versus two dimensional labeling.

5 Space Time Signaling

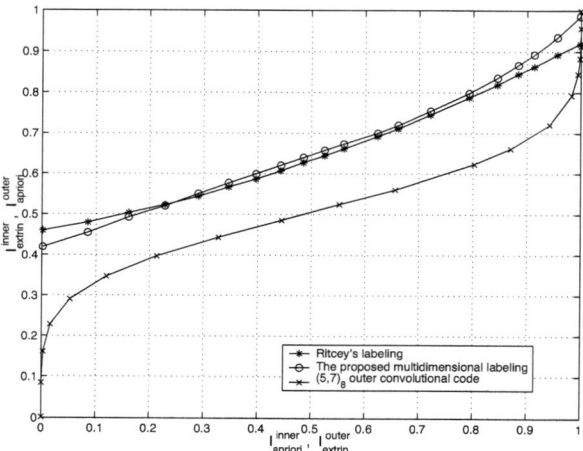

Figure 5.9: EXIT chart of BISTCM-ID system at E_b/N_0 = 6 dB.

Chapter 6

The Multiuser Downlink Scenario (Coding Perspective)

In information theory the broadcast channel is the model where a single transmitter sends individual information to many receivers. In fact this is the case in the downlink of a wireless communication system. However, in existing systems multiple access methods are used in the downlink. In this chapter we will describe the differences of the two possible strategies and show the gain obtained by using broadcast instead of frequency division multiple access for the downlink. Furthermore, we will calculate the mutual information for both cases and show simulation results for Rayleigh fading channels. In addition we describe the benefits using broadcast from a channel coding perspective.

Already Shannon has defined the broadcast channel as a model that one transmitter wants to send individual information to many receivers. Clearly each receiver has a different channel. In [69] the broadcast channel was analysed and it was shown that the capacity region exceeds the one for the multiple access channel. Time division multiple access was assumed. An overview of the results of information theoretic results until 1976 for different multiuser channels can be found in [70].

Figure 6.1 shows the broadcast case for two users. One and the same signal is send to both users which contains individual information for both. Because the signal passes through different channels each user receives a different disturbed signal. From this he detects his individual information. Note, that from an information theoretic perspective the downlink of a wireless communication system is exactly the broadcast case. However, in existing systems this fact is not utilized and a multiple access scheme is applied instead.

In order to illustrate the essential difference between broadcast and multiple access channels we look for example at UMTS (Universal Mobile Telecommunications System). This standard is based on code division multiple access (CDMA). The uplink, from the mobiles to the base station, is a multiple access channel in the sense of information theory because it consists of many transmitters and one receiver. The users transmit at the same time, in the same frequency band, and are separated by codes. In the downlink there is one transmitter and many receivers but also here the users are separated with codes. Thus multiple access is used for the downlink. However, the signal is constructed by adding the spreaded information signals together and transmitting the sum signal. Clearly one single signal is transmitted to all users where the individual signals of all users are simultaneously included. This fact is the same as for the broadcast case. Indeed a receiver has two possibilities. The first is to correlate the received (sum) signal with his individual spreading sequence. Since it is CDMA which is a multiple access scheme the individual information can be extracted. This is called single user detection. The second possibility is to perform

6 The Multiuser Downlink Scenario (Coding Perspective)

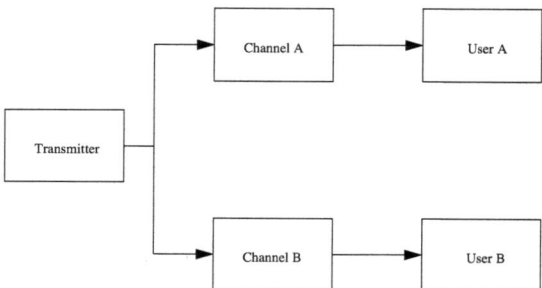

Figure 6.1: Broadcast Transmission (Downlink).

joint detection which means that a receiver does not only detect his own data but all data contained in the sum signal. But this is the broadcast case. And it is commonly known that the performance is much better in case of joint detection than for single user detection [8].

This chapter is structured as follows. In the first section we will compare the broadcast case with frequency division multiple access (FDMA) for two users. Then we will calculate the mutual information between sender and receiver for both cases. We will also discuss the consequences for the use of channel coding in broadcast.

6.1 Comparison of Broadcast and FDMA

We assume that there exist two frequency bands f_1 and f_2, one transmitter, and two users A and B. The four Rayleigh channels shown in Figure 6.2 are assumed to have statistically independent fading coefficients, denoted by h_{1A}, h_{2A}, h_{1B} and h_{2B}. According to [48], the Rayleigh channel can be described with the parameter s_0 and its expectation value $\bar{s}_o$

$$s_0 = ||h||^2 \frac{E_b}{N_0}, \quad \bar{s}_o = E\{s_0\},$$

where $h \in \{h_{1A}, h_{2A}, h_{1B}, h_{2B}\}$.

For FDMA with no channel state information at the transmitter each user gets one of the two frequency bands. Assuming BPSK transmission we can calculate the bit error probability P_A for user A and P_B for B respectively [48] as

$$P_A = P_B = P_b = \frac{1}{2}\left(1 - \sqrt{\frac{\bar{s}_o}{1+\bar{s}_o}}\right).$$

The result is shown in Figure 6.3.

Now we assume that we know all 4 channels at the transmitter. Then we can decide which frequency band we assign to which user before transmission. We select the bands such that in average the smallest bit error rate $P_A + P_B$ is obtained. This curve is shown also in Figure 6.3 and the gain is 4dB at $P_b = 10^{-2}$. Again this curve can be calculated [48] as

6.2 Mutual Information and Channel Coding

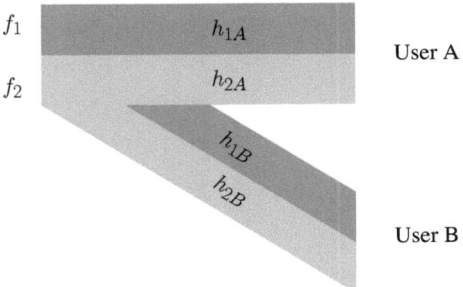

Figure 6.2: FDMA/Broadcast Channels.

$$P_{fr} = \int_0^\infty \frac{P(s_o) \cdot e^{-s_o/\bar{s}_o}}{\bar{s}_o} ds_o \quad \text{with } f \in \{1,2\}, r \in \{A,B\}$$

$$P_{opt,Bpsk} = \min\{P_{1A} + P_{2B}, P_{1B} + P_{2A}\}.$$

Note that this represents an academic case since the signaling of which band is used for which user is completely ignored.

In order to use broadcast and have comparable transmission rates we transmit in both frequency bands the same QPSK symbol. We use the same energy as in the FDMA case and it is equally distributed between the two frequency bands. Each user detects the same QPSK symbol which is received in two bands and takes one of the two bits as his information. Thus one bit is assigned to user A and the other to user B. There is no knowledge of the channel at the transmitter. Figure 6.3 shows that there is a significant gain even compared to the case where the transmitter has channel knowledge. Again this case can be calculated

$$P_{broadcast} = \frac{1}{2}\left(1 - \sqrt{\frac{\bar{s}_o}{2+\bar{s}_o}} \cdot \left(1 + \frac{1}{2+\bar{s}_o}\right)\right),$$

The gain of broadcast is 5 dB at $P_b = 10^{-3}$ compared to FDMA with channel knowledge. Compared to FDMA without channel knowledge is about 6 dB at $P_b = 10^{-2}$. In addition the curve for broadcast has a larger slope which means that the gains become even larger for lower bit error rates.

The results are also valid for higher order modulation alphabets. For example if we use QPSK for FDMA and 16-QAM for broadcast the gain is shown in Figure 6.4. This curves where simulated and different coding techniques for the two 16-QAM symbols are applied (3 curves for broadcast case).

6.2 Mutual Information and Channel Coding

In [71] the authors computed bounds for the mutual information $I(X;Y,Z)$ when a symbol X is transmitted over two channels with conditional probabilities $P(Y|X)$ and $P(Z|X)$. The mutual information is given by

$$I(X;Y,Z) = I(X;Y) + I(X;Z) - I(Y;Z).$$

6   The Multiuser Downlink Scenario   (Coding Perspective)

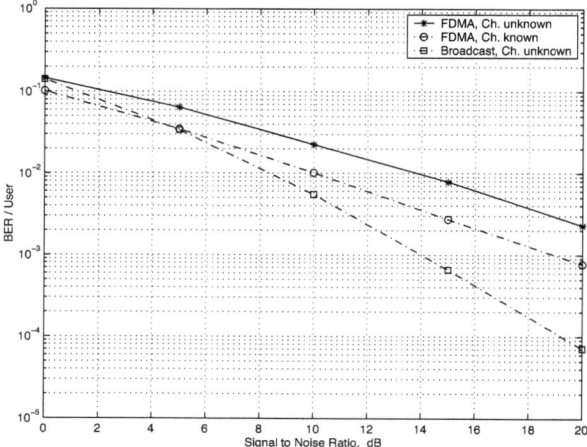

Figure 6.3: Bit error rates for FDMA, FDMA (known channel) and broadcast.

where
$$I(X;Y) = H(X) + H(Y) - H(XY).$$

The function H is the entropy which is given for the the binary symmetric channel with error probability p by
$$H(X) = -p\,\mathrm{ld}(p) - (1-p)\,\mathrm{ld}(1-p) = h(p)$$

Figure 6.5 shows the calculation of $I(X;Y,Z)$ for the broadcast and FDMA case. Thereby the same assumption as in the previous section have been made, namely independent Rayleigh fading, and BPSK for the FDMA and QPSK for the broadcast case. This theoretical considerations demostrate the potential of the broadcast scheme. At a transmission rate of 3/4 bit per symbol we observe a significant gain of more than 4 dB. The mutual information shows also a gain for higher order modulations.

The broadcast description allows to adapt channel coding techniques [19] in order to improve the performance. Since both users receive the same signal they can be encoded by one single codeword of length $2n$ instead of two codewords of length n. In coding theory the codeword error probability P_B can be bounded with the help of the so called Gallager exponent $E(R)$ which depends only on the code rate R by
$$P_B \leq e^{-nE(R)}.$$

This means that as long as the rate is smaller than the capacity ($R < C$) the codeword error probability is exponentially decreasing with the code length n. Since for L users we can apply a code of length $L \cdot n$ in the broadcast case we have an error probability which is smaller by a factor of $1/e^L$ compared with FDMA. However, the transmission rate is the same in both cases because the number of transmitted symbols per time unit is the same.

6.2 Mutual Information and Channel Coding

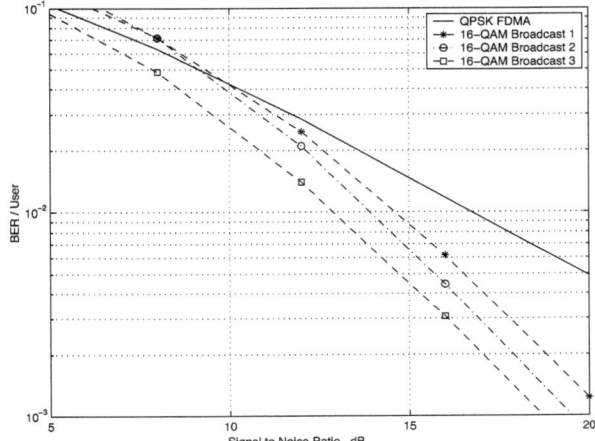

Figure 6.4: Bit error rates for FDMA and broadcast.

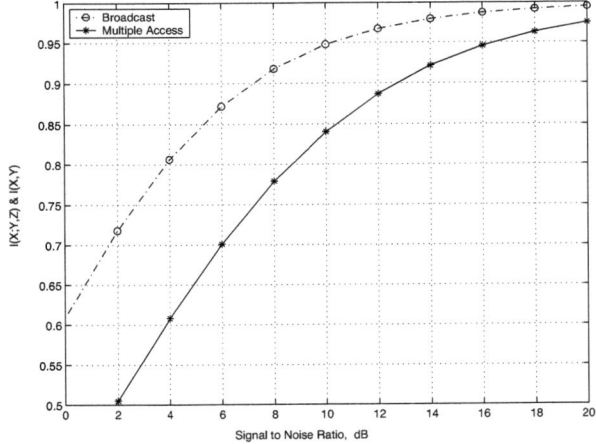

Figure 6.5: Mutual information for FDMA and broadcast.

Chapter 7

Summary and Conclusions

The thesis treated the problem of reliable digital communication over wireless channels when the transmitted signal is subject to multipath propagation. Reliable communication accounts for both improving the reliability of the signal detection, and increasing the transmission rate. Based on the combination of three well known transmission schemes, which are code concatenation, coded modulation and MIMO transmission, we considered three different problems within this general framework. In addition we considered the multiuser case of the downlink transmission.

First in chapter 4, we considered the multidimensional labeling for BICM-ID. MD-BICM-ID results in a low error floor when the signal mapping is optimized. However, the optimized mapping suffers from performance degradation at low SNR regions. By applying modulation doping to the MD-BICM-ID shows that it outperforms BICM-ID in both low and high SNR regions. An interleaver is used to compensate for the degradation on the error floor region introduced by the gray mapping of the symbols. This results in a transmitter which outperforms BICM-ID in the whole SNR region, and MD-BICM-ID at the low SNR regions without additional decoding complexity. We considered 8-PSK modulation to demonstrate this transmission technique.

In chapter 5, we proposed an improvement to the BI-STCM-ID system by considering two consecutive symbols as single entity for designing the mapping rule. The RTS algorithm is applied to obtain an optimized multidimensional 16-QAM signal labeling. The same idea could be applied to obtain multidimensional labeling for other modulation formats, e.g. QPSK and 8-PSK. The performance degradation at the waterfall region is expected and can be improved by using a convolutional code that matches the labeling. EXIT chart is a useful tool for finding the best match between two entities that interact iteratively if the block length is large enough.

A simple transmitter structure for a $2 \times 2 \times 2$ STF transmission based on real rotation of constellation points and orthogonal STC presented in chapter 5. The performance of the optimized STF scheme is shown to outperform a system with a complex rotation matrix. The use of real rotations simplifies the design of the transmitter and reduces the complexity of the decoder. A generalization for a higher frequency diversity order is straight forward. However, parameterizing the real rotation matrix and the search for the optimum parameters is difficult.

In chapter 6, we considered the multiuser case and we have shown that using broadcast for the downlink of wireless communication systems yield significant gains compared to using multiple access schemes. An additional gain can be obtained for broadcast by using one long channel code for all users together instead of many short codes. Moreover, since the encoding for broadcast is not restricted we can always use such encoding which represents CDMA, TDMA, or FDMA. Thus, multiple access can be considered as special case of broadcast encoding. Therefore broadcast can have at least the same or even a better performance than multiple access.

The work in this thesis is far from being complete. A common problem all research students face is to fit his/her research activities into fixed time frame, and still to achieve the original research goals. In this work there are many areas that should be investigated further, and there are other areas that could be extended. The combination of the $2 \times 2 \times 2$ STF transmission with an outer channel code to construct a Multidimensional Bit Interleaved Space Time Frequency Coded Modulation with Iterative Decoding (MD-BISTFCM-ID) is a possible extension to this work. However, the soft output decoding complexity of the multidimensional signals in time, space and frequency should be considered. Also the degradation in performance when transmitting over flat fading channels for the systems considered in this thesis should be studied.

Bibliography

[1] C. E. Shannon, "A Mathimatical Theory of Communication," *Bell Syst. Tech. J.*, vol.27, pp. 379-423 and 623-656, 1948.

[2] I. E. Telatar, "Capacity of Multi-Antenna Gaussian Channels," *European Trans. on Telec.*, vol.10, pp. 585-595, Nov-Dec 1999.

[3] S. Alamouti, "A Simple Transmitter Diversity Scheme for Wireless Communications," *IEEE J. Select. Areas Commun.*, pp. 1451-1458, October. 1998.

[4] Y. Huang, J. A. Ritcey, "Improved 16-QAM Constellation Labeling for BI-STCM-ID with the Alamouti Scheme," *IEEE Commun. Lett.*, vol. 9, no. 2, pp. 157-159, Feb. 2005.

[5] Hagenauer, J., "The Turbo Principle - Tutorial Introduction and State of the Art," *In: Proc. International Symposium on Turbo-Codes*, Brest, France, September 1997.

[6] A. S. Mohammed, M. Bossert, "A Simple $2 \times 2 \times 2$ Full-Rate Full-Diversity Space-Time-Frequency Transmission Scheme," *9th Int. OFDM-Workshop*, Dresden, Germany, Sep. 2004, pp. 275-278.

[7] A. S. Mohammed, W. Hidayat, and M. Bossert. "Multidimensional 16-qam Constellation Labeling of BI-STCM-ID with the Alamouti Scheme," *In Proc. IEEE WCNC 2006*, Las Vegas, USA, April 2006.

[8] M. Bossert and A. S. Mohammed. "Downlink transmission as broadcast channel.," *In Proc. International Symposium on Information Theory and its Applications*, pages 116-119, Parma, Italy, October 2004.

[9] A. S. Mohammed, Yongxiang Gong and M. Bossert. "On Multidimensional BICM-ID with 8-PSK Constellation Labeling," *In Proc. PIMRC 2007*, Athens, Greece, Sept. 2007.

[10] M. Borran, M. Memarzadeh, B. Aazhang, "Design of Coded Modulation Schemes for Orthogonal Transmit Diversity," *ISIT*, Washington, D.C., June 2001, pp. 339-344.

[11] Forney, G. D., Jr., "Concatenated Codes," *Cambridge, MA: MIT Press*, 1966.

[12] Yuen, J. H., et al., "Modulation and Coding for Satellite and Space Communications," *Proc. IEEE*, vol. 78, no. 7, July 1990, pp. 1250-1265.

[13] N. Seshadri, J. H. Winters, "Two signaling schemes for improving the error performance of frequency-division-duplex (FDD) transmission systems using transmitter antenna diversity," *Int. J.Wireless Inform. Networks*, vol. 1, no. 1, pp. 49-59, 1994.

[14] T. Muharemovic and B. Aazhang, "Information Theoretic Optimality of Orthogonal Space-Time Codes and Concatenated Code Construction," *Int. Conf. on Telec.*, June 2000.

[15] Nghi. H. Tran, Ha. H. Nguyen, "Signal Mappings of 8-ary Constellations for BICM-ID over a Rayleigh Fading Channel," *Elect. and Comp. Eng., 2004 Canadian Conference*, Vol.3 pp. 1809-1813, 2-5 May 2004.

[16] N. H. Tran, H. H. Nguyen, "Improving the Performance of QPSK BICM-ID by Mapping on the Hypercube," *Proc. of IEEE VTC*, Sept. 2004.

[17] R. Battiti, G. Tecchiolli, "Basic Reactive Tabu Search Software Routines and Specific Routines for the Quadratic Assignment Problem," *Available online:* http://rtm.science.unitn.it/~battiti/reactive.html.

[18] R. Battiti, M. Brunato, The Reactive Search Website, [Online]: http://www.reactive-search.org/.

[19] M. Bossert, *Channel Coding for Telecommunications*, John Wiley & Sons, Ltd., 1999, ISBN 0-471-98277-6.

[20] E. Zehavi, "8-PSK Trellis Codes for a Rayleigh Channel," *IEEE Trans. Commun.*, vol. 40, no. 5, pp. 873-844, May 1992.

[21] G. Caire, G. Toricco, E. Biglieri, "Bit-Interleaved Coded Modulation," *IEEE Trans. Inform Theory*, vol. 44, no. 3, pp. 927-946, May 1998.

[22] X. Li, J. A. Ritcey, "Bit-Interleaved Coded Modulation with Iterative Decoding," *IEEE Commun. Letters*, vol.1, no. 6, pp. 169-171, November 1997.

[23] X. Li, J. A. Ritcey, "Trellis-Coded Modulation with Bit interleaving and Iterative Decoding ," *IEEE Journal on Selected Areas in Commun.*, vol.17, no. 4, pp. 715-724, April 1999.

[24] S. Dolinar and D. Divsalar, "Weight Distributions for Turbo Codes Using Random and Nonrandom Permutations," *TDA Progress Report* 42-122, August 15, 1995.

[25] X. Li, A. Chindapol, J. A. Ritcey, "Bit-Interleaved Coded Modulation with Iterative Decoding and 8PSK Signaling," *IEEE Trans. Commun.*, vol. 50, no. 8, pp. 1250-1257, August 2002.

[26] A. Chindapol, J. A. Ritcey, "Design, Analysis, and Performance Evaluation for BICM-ID with Square QAM Constellations in Rayleigh Fading Channels," *IEEE Journal on Selected Areas in Commun.*, vol. 19, no. 5, pp. 944-957, May 2001

[27] L.-F. Wei, "Trellis-Coded Modulation with multidimensional constellation," *IEEE Trans. Inform. Theory*, vol. 33, pp. 483-501, July 1987.

[28] S. S. Pietrobon, R. H. Deng, A. Lafanechere, G. Ungerboeck, D. J. Castello, "Trellis-Coded Multi-dimensional Phase Modulation," *IEEE Trans. Inform. Theory*, vol. 36, no. 1, pp. 62-89, Jan. 1990.

[29] F. Simoens, H. Wymeersch, H. Bruneel, M. Moeneclaey, "Multi-dimensional Mapping for Bit-Interleaved Coded Modulation with BPSK/QPSK Signaling," *IEEE Communications Letters*, vol. 9, no. 5, pp. 453-455, May 2005.

[30] L. Szczecinski, H. Chafnaji, and C. Hermosilla, "Modulation Doping for Iterative Demapping of Bit-Interleaved Coded Modulation," *IEEE Communications Letters*, vol. 9, no. 12, Dec. 2005

[31] F. Schreckenbach, N. Goertz, J. Hagenauer, and G. Bauch, "Optimization of Symbol Mappings for Bit-Interleaved Coded Modulation With Iterative Decoding," *IEEE Communications Letters*, vol. 7, no. 12, Dec. 2003.

[32] F. Schreckenbach, G. Bauch, "Irregular Signal Constellations, Mappings and Precoder," in *Proc. International Symposium on Information and its Applications (ISITA)*, Oct. 2004.

[33] G. D. Golden, C. J. Foschini, R. A. Valenzuela, and P. W. Wolniansky, "Detection algorithm and initial laboratory results using V-BLAST space-time communications architecture," *Elect. Lett.*, vol. 35, pp. 14-16, Jan. 1999.

[34] S. Lin and D. J. Costello, "Error Control Coding," *Prentice-Hall, Inc. Englewood Cliffs*, New Jersey 07632, 1983.

[35] J. L. Massey, "Coding and Modulation in Digital Communications," *Proc. of International Zurich Seminar on Digital Communications*, March 1974.

[36] Gottfried Ungerboeck, Channel Coding with Multilevel/Phase Signals, *IEEE Trans. Inform. Theory*, vol. IT-28, No. 1, pp. 55-66,January 1982.

[37] Gottfried Ungerboeck, Trellis-Coded Modulation with Redundant Signal Sets, Part I: Introduction, *IEEE Commun. Mag.*, vol. 25, pp. 5-11, Feb. 1987.

[38] D. Disvalar and M. K. Simon, The Design of Trellis Coded MPSK for Fading Channels: Performance Criteria, *IEEE Trans. Commun.*, vol. 36, pp. 1004-1011, Sept. 1988.

[39] C. Berrou, A. Glavieux, P. Thitimajshima, Near Shannon Limit Error Correction Coding and Decoding: Turbo Codes, *Proc. IEEE Int. Conf. Commun.*,pp. 1064-1070,1993

[40] Yuheng Huang, Bit-Interleaved Coded Modulation with Iterative Decoding for Wireless Communications, *Doctor of Philosophy Dissertation*, University of Washington, 2004.

[41] J. Winters, On the Capacity of Radio Communication Systems with Diversity in a Rayleigh Fading Environment, *IEEE Journal on Selected Areas in Commun.*, vol. 5, pp. 871-878, June 1987.

[42] V. Tarokh, N. Seshadri, A. R. Calderbank, Space-Time Codes for High Data Rate Wireless Communication : Performance Criterion and Code Construction, *IEEE Trans. Inform. Theory*, vol. 44, no. 2, pp. 744-765, March 1998.

[43] V. Tarokh, H. Jafarkhani, A. R. Calderbank, Space-Time Block Codes from Orthogonal Designs, *IEEE Trans. Inform. Theory*, vol 45, no. 5, pp. 1456-1467, July 1999.

[44] V. Tarokh, H. Jafarkhani, A. R. Calderbank, Space-Time Block Coding for Wireless Communications : Performance Results, *IEEE Journal on Selected Areas in Commun.*, vol 17, no. 3, pp. 451-460, March 1999.

[45] R. K. Ahuja, James B. Orlin, Ashish Tiwari, A Greedy Genetic Algorithm for the Quadratic Assignment Problem, *Computers and Operations Research* vol 27, pp. 917-934.

[46] R. Battiti, G. Tecchiolli, The Reactive Tabu Search, *ORSA Journal on Computing*, vol. 6, no. 2, pp. 126-140, 1994.

[47] A. J. Viterbi, J. K. Wolf, E. Zehavi, R. Padovani, A Pragmatic Approach to Trellis-Coded Modulation, *IEEE Commun. Mag.*, vol. 27, pp. 11-19, July 1989.

[48] J. G. Proakis, *Digital Communications*, McGraw-Hill, 2001.

[49] S. Benedetto, D. Divsalar, G. Montorsi, F. Pollara, A Soft-Input Soft-Output APP Module for Iterative Decoding of Concatenated Codes, *IEEE Commun. Letters*, vol. 47, no. 1, pp. 22-24, Jan. 1997.

[50] S. Benedetto, D. Divsalar, G. Montorsi, F. Pollara, Serial Concatenation of interleaved codes: Performance analysis, design and iterative decoding, *IEEE Transactions on information Theory*, vol. 44, no. 3, pp. 909-926 , May. 1998.

[51] L. R. Bahl, J. Cocke, F. Jelinek, and J. Raviv, Optimal Decoding of Linear Codes for Minmizing Symbol Error Rate, *IEEE Transactions on information Theory*, pp. 284-287 , Mar. 1974.

[52] Nghi H. Tran, Ha. H. Nguyen, Signal Mappings of 8-ary Constellations for BICM-ID over a Rayleigh Fading Channel, *Electr. and Comp. Eng. 2004 Canadian Conference*, vol.3, pp. 1809-1813, 2-5 May 2004.

[53] E. Biglieri, G. Caire, G. Tarrico, J. Ventura, Simple Method for Evaluating Error Probabilities, *Electron. Lett.*, vol. 32, no. 3, pp. 191-192, Feb. 1996.

[54] H. Schulze, System Design for Bit Interleaved Coded QAM with Iterative Decoding in a Rician Fading Channel, *European Transactions on Telecommunications*, vol. 14, pp.119-129, April 2003.

[55] M. K. Simon, D. Divsalar, Some New Twist to Problems Involving the Gaussian Probability Integral, *IEEE Trans. Commun.*, vol. 46, pp. 200-210, Feb. 1998

[56] G. J. Foschini, Layered Space-Time Architecture for Wireless Communication in a Fading Environment when Using Multiple Antennas, *Bell Labs. Tech. Journal*, vol. 1, pp. 41-59, Autumn 1996.

[57] Z. Hong, B. L. Hughes, Bit-Interleaved Space-Time Coded Modulation with Iterative Decoding, *IEEE. Trans. Wireless Commun.*, vol. 3, no. 6, pp. 1912-1917, November 2004.

[58] Yuheng Huang, James A. Ritcey, Tight BER Bounds for Iteratively Decoded Bit-Interleaved Space-Time Coded Modulation, *IEEE Commun. Letters*, vol. 8, no. 3, pp. 153-155, March 2004.

[59] Stephan ten Brink, Design of Serially Concatenated Codes based on Iterative Decoding Convergence, *2nd Symposium on Turbo Codes*, Brest France, Sept. 2000.

[60] Lutz H. -J. Lampe, Robert Schober, Robert F. H. Fischer, Multilevel Coding for Multiple-Antenna Transmission, *IEEE Trans. Wireless Commun.*, vol. 3, no. 1, pp. 203-208, Jan. 2004.

[61] Joseph Boutros, Nocolas Gresset, L. Brunel, Turbo Coding and Decoding for Multiple Antenna Channels, *International Symposium on Turbo Codes and Related Topic*, Brest, Sept. 2003.

[62] A. Hübner, *On Permutor Design Aspects for Concatenated Convolutional Codes*, Fortschritt-Berichte VDI, Reihe 10, Nr. 749, Doctoral Thesis, University of Ulm, Germany, 2004, ISBN 3-18-374910-6.

[63] B. Hassibi and B. M. Hochwald, High-Rate Codes that are Linear in Space and Time, *IEEE Trans. on Info. Theory*, vol. 48, no. 7, pp. 1804-1824, July 2002.

[64] Z. Liu, Y. Xin and G. B. Giannakis, Space-Time-Frequency Coded OFDM over Frequency-Selective Fading Channels, *in IEEE Trans. on Signal Proc.* vol. 50, No. 10, Oct. 2002, pp. 2465-2476.

Bibliography

[65] Andreas F. Moisch, Moe Z. Win, and Jack H. Winters, Space-Time-Frequency (STF) Coding for MIMO-OFDM Systems, *IEEE Commun. Lett.*, vol. 6, NO. 9, pp. 370-372, Sept 2002.

[66] J. Boutros and E. Viterbo, Signal Space Diversity: A Power and Bandwidth Efficient Diversity Technique for the Rayleigh Fading Channel, *IEEE Trans. Inform. Theory*, vol. 44, pp. 1453-1467, July 1998.

[67] E. Viterbo and J. Boutros, A Universal Lattice Code Decoder for Fading Channels, *IEEE Trans. Inform. Theory*, vol. 45, NO. 5, pp. 1639-1642, July 1999.

[68] M. O. Damen, A. Chkeif, and J. C. Belfiore, Lattice Code Decoder for Space-Time Codes, *IEEE Commun. Lett.*, vol. 4, pp. 161-163, May 2000.

[69] T. M. Cover, Broadcast channels, *IEEE Trans. Inf. Theory*, IT-18:214, January 1972.

[70] E. C. van der Meulen, A survey of multi-way channels in information theory: 19611976, *IEEE Trans. Inf. Theory*, IT-23(1):137, January 1977.

[71] I. Land, P. Hoeher, S. Huettinger, and J. Huber, Bounds on information combining, *In Proc. 3rd Int. Symp. on Turbo Codes & Related Topics*, pages 3942, Brest, France, 2003.

Die VDM Verlagsservicegesellschaft sucht für wissenschaftliche Verlage abgeschlossene und herausragende

Dissertationen, Habilitationen, Diplomarbeiten, Master Theses, Magisterarbeiten usw.

für die kostenlose Publikation als Fachbuch.

Sie verfügen über eine Arbeit, die hohen inhaltlichen und formalen Ansprüchen genügt, und haben Interesse an einer honorarvergüteten Publikation?

Dann senden Sie bitte erste Informationen über sich und Ihre Arbeit per Email an *info@vdm-vsg.de*.

Sie erhalten kurzfristig unser Feedback!

VDM Verlagsservicegesellschaft mbH
Dudweiler Landstr. 99
D - 66123 Saarbrücken

Telefon +49 681 3720 174
Fax +49 681 3720 1749

www.vdm-vsg.de

Die VDM Verlagsservicegesellschaft mbH vertritt

Printed by Books on Demand GmbH, Norderstedt / Germany